Routledge Revivals

Mineral Wealth and Economic Development

Over the last several decades, many low-income mineral exporting countries have seen their per capita income decline or their standard of living stagnate. In this title, prominent analysts identify reasons behind the distressing economic performance of these countries including ineffective public policies, political misuse of mineral rents, and the deleterious effects of economic nationalism on the foreign investment climate in developing countries. Originally published in 1992, this title remains relevant for students interested in environmental studies and public policy.

Part One

Mineral Wealth and Economic Development

Mineral Wealth and Economic Development

Edited by
John E. Tilton

RFF PRESS
RESOURCES FOR THE FUTURE

First published in 1992
by Resources for the Future, Inc.

This edition first published in 2016 by Routledge
2 Park Square, Milton Park, Abingdon, Oxon, OX14 4RN
and by Routledge
711 Third Avenue, New York, NY 10017

Routledge is an imprint of the Taylor & Francis Group, an informa business

Publisher's Note
The publisher has gone to great lengths to ensure the quality of this reprint but
points out that some imperfections in the original copies may be apparent.

Disclaimer
The publisher has made every effort to trace copyright holders and welcomes
correspondence from those they have been unable to contact.

A Library of Congress record exists under LC control number: 91046042

ISBN 13: 978-1-138-19302-4 (hbk)
ISBN 13: 978-1-315-63933-8 (ebk)

MINERAL WEALTH
AND ECONOMIC DEVELOPMENT

edited by JOHN E. TILTON

John M. Olin Distinguished Lectures
in Mineral Economics

Printed in the United States of America

Published by Resources for the Future
1616 P Street, NW; Washington, DC 20036-1400

Books from Resources for the Future are distributed worldwide by
The Johns Hopkins University Press.

Library of Congress Cataloging-in-Publication Data

Mineral wealth and economic development / John E. Tilton, editor.
    p.   cm.
    "John M. Olin distinguished lectures in mineral economics."
    Includes bibliographical references.
    ISBN 0-915707-62-4 (alk. paper)
    1. Mineral industries—Congresses.   2. Mines and mineral resources—Congresses.
  3. Balance of payments—Congresses.   4. Economic development—Congresses.
  5. Mineral industries—Developing countries—Congresses. 6. Mines and mineral resources—Developing countries—Congresses. 7. Balance of payments—Developing countries—Congresses. I. Tilton, John E.
HD9506.A2M5433   1992
338.2—dc20   91-46042
                                                         CIP

The paper in this book meets the guidelines for permanence and durability of the Committee on Production Guidelines for Book Longevity of the Council on Library Resources.

The book is the product of the Energy and Natural Resources Division at Resources for the Future, Douglas R. Bohi, director. It was edited by Nancy A. Winchester and designed by Brigitte Coulton. The cover was designed by Gehle Design.

Contributors

Olivier Bomsel

Associate Director, Centre d'Economie des Ressources Naturelles, Ecole Nationale Supérieure des Mines, Paris, France

Philip Daniel

Fellow, Institute of Development Studies, University of Sussex, Brighton, United Kingdom

Theodore H. Moran

Karl F. Landegger Professor and Director, Program in International Business Diplomacy, School of Foreign Service, Georgetown University, Washington, D.C.

Marian Radetzki

Director, SNS Energy, Stockholm, Sweden, and Visiting Professor, Mineral Economics Department, Colorado School of Mines, Golden, Colorado

John E. Tilton

William J. Coulter Professor and Head, Mineral Economics Department, Colorado School of Mines, Golden, Colorado, and University Fellow, Resources for the Future, Washington, D.C.

Contents

Economic Development and the Timing of Mineral Exploitation **39**

Marian Radetzki

The Political Economy of Rent in Mining Countries **59**

Olivier Bomsel

Economic Policy in Mineral-Exporting Countries: What Have We Learned? **81**

Philip Daniel

Preface

The relationship between mineral wealth and economic development, particularly in the low-income, mineral-exporting countries, is perplexing. Many of these countries have suffered a drastic decline in per capita income over the past several decades. Others, only slightly more fortunate, have watched their standard of living stagnate.

Such poor economic performance raises a number of questions. Is it possible that domestic mineral wealth actually retards growth and development? If so, is this perverse outcome inevitable? What advice, if any, can mineral economists offer public officials striving to turn their country's mineral resources into assets rather than liabilities in the war against poverty and deprivation?

In the hope of providing answers to these questions, the Colorado School of Mines selected Mineral Wealth and Economic Development as the theme for its 1989 John M. Olin Distinguished Lectureship Series in Mineral Economics. Each year, under the lectureship program, the Mineral Economics Department invites several well-known economists from industry, government, and academia to spend several days in residence with students and faculty. During their visits, participants present a public lecture on a topic of special interest to them within the theme of the lectureship series.

The papers in this volume, with the exception of the Overview, were originally prepared for the Olin Lectureship Series on Mineral Wealth and Economic Development and subsequently revised in light of comments and suggestions. The Colorado School of Mines is grateful to the John M. Olin Foundation for generously providing the financial support for the lectureship series and to Resources for the Future (RFF) for making this material available to a broad audience. We would also like to thank Lita G. Dunham and Nancy A. Winchester for carefully editing the volume, and Dorothy Sawicki and Brigitte Coulton at RFF for shepherding it through the editing and production process.

John E. Tilton
Coulter Professor and Head,
    Mineral Economics Department
Colorado School of Mines

Mineral Wealth and Economic Development: An Overview

JOHN E. TILTON

C ommon sense suggests that mineral wealth, like other assets, should help countries develop and grow. After all, the returns from mineral exploitation can be used to build airports and highways, stores and factories, schools and hospitals, and homes and parks. They can enhance political stability by addressing regional or tribal grievances and in various other ways bolster economic growth.

Mining and mineral processing can also generate jobs, provide opportunities for the development of domestic skills, encourage the creation of associated industries, and provide other beneficial side effects or linkages for the local economy.

History documents that mineral resources can indeed facilitate economic development. The Industrial Revolution began in England and quickly spread to Germany and the United States partly because these countries were well endowed with coal and other natural resources. Saudi Arabia and other Middle East oil-producing countries are more recent examples of the positive role mineral wealth can have in economic development.

Thus, the recent poor economic performance of so many low-income mineral-exporting countries is both surprising and distressing. Many of them have watched their standard of living deteriorate relative to developing countries less blessed with mineral resources. A number have even suffered from complete economic stagnation and decline.

The depressed conditions that plagued most of the nonfuel mineral markets from the mid-1970s to the mid-1980s are only partly to blame. Growth in the less developed mineral-exporting countries remains an enigma, suggesting that the relationship between mineral wealth and economic development is far more complex and at times far less favorable than common sense would suggest.

Three critical considerations influence the extent to which mineral resources promote or inhibit economic development. First, so long as mineral wealth remains in the ground, it is a dormant asset. Without exploration a country may not even realize it possesses valuable deposits, and once discovered, such deposits can contribute to economic growth only if they are exploited.

Second, the returns from the mining and processing of mineral deposits must be invested to enhance the future flow of goods and services. These returns, which economists call rents, depend on the present value of the future stream of net revenues that mineral deposits can generate. (Net revenues are defined in the economic sense as the difference between total revenues and total costs, where costs cover a competitive rate of return for capital and a competitive rate of compensation for entrepreneurship.)

Rents lost through waste and needlessly high production costs contribute nothing to economic growth. The same holds true for rents spent on current consumption or rents captured and expatriated by foreign interests. Even those rents that are invested can retard economic growth if they are used unwisely.

Third, mining and mineral processing may alter the domestic economy and affect growth in ways that go beyond the generation and use of mineral rents. Mining can create jobs, train workers, and provide other positive side effects. However, these nonrent effects, sometimes referred to as economic linkages, are not always beneficial; they can inhibit as well as encourage economic growth. A resource boom, for instance, may cause a country's currency to appreciate, making it more difficult for its domestic industries to compete in both domestic and foreign markets.

The papers in this volume focus on these three critical considerations. The lectures were originally presented in a public series that had two overriding goals: (1) a better understanding of the complex relationship between mineral wealth and economic development, including the reasons why mineral resources in some low-income countries have apparently retarded rather than fostered growth, and (2) specific policy guidance for governments hoping to use their mineral wealth as a means of economic development.

The authors who address these themes are especially well qualified to do so. Theodore Moran's interest in the role of mining in developing countries goes back to his days as a graduate student in the Department of Government at Harvard University. His dissertation and subsequent book, *Multinational Corporations and the Politics of Dependence: Copper in Chile 1945–1973*, examines the complex and often hostile relationship between the Chilean government and that country's multinational mining companies. Over the years he has maintained a strong interest in political risk and direct foreign investment in developing countries and has published several books and many professional papers in this area. He has advised government agencies and private corporations in both the United States and developing countries. As a member of the Policy Planning Staff of the U.S. Department of State in the late 1970s, he was responsible for North–South economic issues, including international investment in mining and metal processing.

Marian Radetzki received his doctorate from the Stockholm School of Economics and served for several years as the chief economist for CIPEC, the Intergovernmental Council of Copper Exporting Countries. A senior research fellow at the Institute for International Economic Studies at the University of Stockholm for many years, he has published widely in the fields of mineral and development economics. His two most recent books are *State Mineral Enterprises* and *A Guide to Primary Commodities in the World Economy*.

Olivier Bomsel, an Ingenieur des Mines with a doctorate in economics, directs the economic research and consulting work on minerals and materials of the Centre d'Economie des Ressources Naturelles (CERNA) at the Ecole Nationale Supérieure des Mines in Paris. He is well known for his work on the mining industry in the developing countries of Africa. His latest book is entitled *The End of Large Mining Projects in the Third World*.

Philip Daniel, also an economist, served from 1986 to 1989 as Special Adviser in the Technical Assistance Group in the Commonwealth Secretariat in London, advising the governments of developing countries negotiating exploration and production agreements with foreign mineral and petroleum companies. Daniel worked in Papua New Guinea during the late 1970s in what is now the Department of Finance and Planning and served as a member of the country's minerals and petroleum negotiating team. At the Institute of Development Studies at the University of Sussex, he manages the research program on productive enterprise and continues to advise governments and international organizations on minerals issues. He has published two books on development problems in mineral-producing countries: one on Zambia entitled *Africanization, Nationalization and Inequality,* and one with Roderick Sims entitled *Foreign Investment in Papua New Guinea.*

In addressing mineral wealth and economic development, the authors have drawn on their own unique experiences and perspectives. Differences in terminology and denotation occasionally arise, and not all the contributors work from the same body of literature or source material. Nor, given the public venue of their original addresses, do they always provide the scholarly documentation for their sometimes provocative remarks that the reader of academic essays would legitimately expect. Their informed assessments nevertheless provide many unique and valuable insights into the actual and potential contributions of mining and mineral processing in the economic growth of nations.

Economic Nationalism and Multinational Mining Companies

The 1960s and early 1970s were difficult years for multinational mining companies, particularly those operating in developing countries. In some countries, governments nationalized mines and processing facilities. In others, they confiscated all or part of the revenue-earning ability of foreign-owned ventures through higher taxes and other measures. As a result, the multinational mining corporations, which for many years had dominated mining and mineral processing outside the centrally planned countries, withdrew from many parts of the world and concentrated their talents and resources on the United States and a handful of other developed countries.

Since the mid-1970s, the international climate for foreign investment has improved, in part a by-product of the depressed conditions that for so long prevailed in many mineral markets. With prices down, the rents from mining and mineral processing declined or vanished entirely. The capital and technical expertise needed to develop new ventures became increasingly difficult to attract. Moreover, many of the newly independent developing countries were, by the 1980s, politically more mature. Fears of economic imperialism and neocolonialism were on the wane, and economic policy reflected a growing appreciation of the positive contributions of foreign investment.

These developments, along with the growing worldwide infatuation with private enterprise and freer markets, have led many to conclude that the earlier conflict between host governments and foreign mining companies was a unique event, the result of the inexperience of newly independent countries and other special circumstances of the 1960s. According to this view, the 1990s will see the trend toward greater cooperation between governments and foreign mining companies continue, with the companies providing the expertise and capital needed to exploit the mineral resources of the developing world.

Moran scrutinizes this optimistic view in his paper, raising a note of caution. He offers two explanations for the economic nationalism of the 1960s and early 1970s. The first and most commonly encountered encompasses a variety of host country concerns that in Moran's words were "fundamentally nonrational, extraeconomic, ideological, or purely emotional." In short, policymakers and the public simply did not like the domestic mineral sector dominated by foreign interests, regardless of the economic benefits. Moran calls this explanation the nationalism-as-psychic-gratification hypothesis for economic nationalism.

The second explanation, the obsolescing-bargain hypothesis, contends that host government demands for nationalization or for the fundamental renegotiation of mining agreements came about as the risk and uncertainty surrounding successful ventures declined over time. In this case, the objective of the confrontation was, in an economic sense, quite rational. Governments simply were trying to capture a larger share of the rents associated with multinational projects, an objective that became more feasible after foreign companies had made sizable investments and the magnitude of the rents was known with greater certainty.

If psychic gratification was primarily responsible for the deterioration in the foreign investment climate during the 1960s, Moran concedes that attitudes have since changed in many Third World countries and that these concerns are now of much less importance. He suggests, however, that the obsolescing-bargain hypothesis was a strong force behind the confrontations of the 1960s. Similar confrontations could arise in the 1990s in light of the recent recovery of mineral prices and the much larger streams of mineral rents flowing in their wake. This, according to Moran, has "ominous implications for the role that mining may play in Third World development."

Moran's focus is on the first of the three critical considerations: exploitation of mineral wealth. He fears the foreign investment climate in developing countries could deteriorate, leaving many without the capital and expertise needed to exploit their mineral assets. Research over the past decade reveals that certain types of foreign investments are relatively immune to the dangers of economic nationalism. Among them are the following:

- Projects with relatively low capital costs that investors can convincingly threaten to abandon
- Projects whose economic viability requires access to rapidly changing technology
- Projects producing differentiated goods, particularly those where advertising and brand identification are used to create and maintain consumer loyalty
- Projects in highly concentrated industries where the number of producing firms is small.

Mining and mineral-processing ventures with few exceptions meet none of these conditions, forcing Moran to conclude that the mineral sector suffers from a structural vulnerability with respect to economic nationalism. This leads him to examine corporate strategies for reducing political risk, such as developing market control over downstream processing, sharing ownership with the host government through joint ventures, and pursuing an incremental investment policy whereby a project is developed in a series of discrete steps over an extended period. For a number of reasons, he concludes that these strategies do not offer multinational mining companies much protection.

Despite these difficulties, the paper ends on a positive note. The basic problem is the durability of agreements between host governments and foreign investors and the investor's perception of the ability of current governments to commit future governments, particularly when changes in government are likely to take place.

Moran suggests that a solution lies in creative arrangements that involve the use of project financing. Such agreements are structured so that the failure of future governments to comply with their terms provokes retaliation from an international consortium of lenders, purchasers, and others, including the World Bank and other international organizations. This arrangement greatly increases the cost and hence reduces the likelihood of unilateral abrogation or alteration.

Such a proposal might at first appear as an undesirable constraint on the economic sovereignty of independent states. If successful, however, it would enhance the economic authority of current governments by providing them with creditable means of binding future governments, thereby increasing their own ability to negotiate long-term agreements with foreign investors. Moran concludes that such strategies "should allow promising but risky projects to go forward that otherwise would be left on the drawing boards, enhancing greatly the contribution that mining can ultimately make to Third World development."

The Optimal Pace of Mineral Exploitation

While Moran focuses on the climate for foreign investment, Marian Radetzki examines another reason why the mineral wealth of developing countries may remain too long in the ground, a dormant asset contributing nothing to economic development. Specifically, he is concerned that public policy is unduly influenced by arguments favoring the deferral of mineral exploitation.

The analysis begins with Hotelling's theory of resource exhaustion, which predicts that the value of mineral resources left in the ground will rise over time at the rate of interest. Hotelling's work, which dates back to the early 1930s, constitutes a seminal contribution to the theoretical literature in this area. But it assumes a world with no uncertainty, no exploration and discovery, and no new technology. More recent and realistic versions of exhaustion theory find that the value of mineral resources need not rise at the rate of interest. Indeed, under certain conditions these theories even anticipate declining mineral values.

Policymakers and others are fond of discounting the influence of economic theory in the real world, particularly theory based on assumptions as unrealistic as Hotelling's. Yet, most decision makers are influenced by conceptual models or perceptions that are abstractions and simplifications of the real world. These models are molded and shaped far more often than generally recognized by economic theories that are now obsolete. In the words of John Maynard Keynes (1964, pp. 383–384),

> The ideas of economists and political philosophers, both when they are right and when they are wrong, are more powerful than is commonly understood. Indeed the world is ruled by little else. Practical men, who believe themselves to be quite exempt from any intellectual influences, are usually the slaves of some defunct economist. Madmen in authority, who hear voices in the air, are distilling their frenzy from some academic scribbler of a few years back. I am sure that the power of vested interests is vastly exaggerated compared with the gradual encroachment of ideas. . . . It is ideas, not vested interests, which are dangerous for good or evil.

For Radetzki, the ghost of Hotelling still stalks the corridors of government buildings around the world encouraging the widespread belief that resources not used today will be worth more tomorrow.

Indeed, Hotelling provides intellectual support for those who favor postponing mineral exploitation for other reasons, such as the desire to save mineral wealth for future generations or for the day when domestic industries will need these resources. This argument, with its understandable appeal to domestic interests, is much easier to accept if the value of unexploited resources left in the ground is likely to appreciate at roughly the same rate as the potential rents invested at the prevailing rate of interest, as the Hotelling model suggests. However, it overlooks the future increase in mineral supplies flowing from new discoveries and technological change. Perhaps even more serious, technological change can render once-commercial deposits uneconomic. Radetzki notes that the Swedish iron ore industry in recent years has lost its competitiveness throughout much of Europe—its traditional market—because major developments in bulk shipping have lowered the delivered cost of Australian and Brazilian iron ore. Had Sweden decided in the early nineteenth century to save its iron ore, the rents realized by previous generations would have been lost with little or no benefit for future generations. Indeed, to the extent these rents have contributed to the country's growth, future generations would have been worse off without them.

A different rationale for postponing mineral exploitation arises from the desire to avoid sharing the resulting rents with foreign investors. To keep all the rents, a country must have the necessary physical and human capital for mineral exploitation available domestically. Further, there are significant entry costs associated with mining and mineral processing. Because these costs have to be incurred at some time, however, they do not by themselves constitute a valid argument for postponement. In Radetzki's words, "The mineral rent that the foreigners take for the services they provide may appear excessive, but it can also be regarded as a temporary and unavoidable education expenditure."

Radetzki finds two other arguments for deferral equally lacking. The first contends that very large projects, such as the Escondida copper project in Chile, can exert so much downward pressure on world prices that the host government may actually suffer a decline in its mineral profits and foreign exchange earnings. But for this to happen requires three very restrictive assumptions: (1) once on stream, the project must account for a sizable portion of world supply, (2) the host country must previously have contributed a large share of world production, and (3) a decline in price must have little effect on both world supply and demand. This last assumption, while perhaps valid for a year or two, is questionable over the life of major mining projects.

The second argument, which uses OPEC (the Organization of Petroleum Exporting Countries) as an example, highlights the benefits of raising market price through producer cartels and other types of monopolistic measures. To be successful, such efforts require producers to restrict current production or exports. Very few mineral cartels manage to maintain prices above competitive levels for more than a few years. After the cartels collapse, market prices fall and remain depressed for a number of years. So once again, a relatively rare combination of conditions is required to make deferral of mineral exploitation worthwhile.

Radetzki does not claim that host governments should never postpone the development of profitable mineral deposits, but he argues that too often the potential contribution of mineral wealth to economic development is lost or dissipated as a result of mistaken decisions to proceed too slowly or not at all. In his words, "in most cases . . . a developing country would be best off by making use, without delay, of any profitable mineral resource potential it possesses."

Political Pressures and the Use of Mineral Rents

Olivier Bomsel turns to the second condition affecting the contribution of mining to economic development, the use of mineral rents. He concentrates on 15 low-income countries whose nonfuel mineral exports have accounted for over 40 percent of total merchandise exports at least once since 1975. For the most part, these countries are poor and depend heavily on mining for government revenues as well as foreign exchange. As a result, their mineral wealth can play a particularly favorable role in their economic development.

Despite this potential, the economies of these countries have stagnated in recent years. In 13, per capita gross domestic product actually fell between 1975 and 1987, sometimes by more than 30 percent. The depressed mineral markets over much of this period accounted for part of this poor performance, but the perverse use of mineral rents seriously aggravated the situation.

To stimulate economic growth, rents must be invested in education, infrastructure, industry, and other activities that raise future income. The mining countries Bomsel examines rarely use their mineral rents in this virtuous manner. Instead a vicious cycle tends to arise, caused by what Bomsel calls the political economy of mineral rents, that actually inhibits economic development.

Rents are used to offset the painful disruptions associated with structural changes elsewhere in the economy. Food, for example, may be imported to supplement declining domestic production and thereby keep domestic prices from rising. Although this helps consumers maintain their standard of living, the underlying problems of the agricultural sector are ignored. As this sector continues to decline, more and more imports are needed. Unemployed peasants move to urban areas, creating pressure for jobs. The government responds by increasing employment in the public sector and state-run enterprises. As the country becomes increasingly dependent on mineral rents to support such unproductive activities, it squeezes the mineral sector for more. This scares off foreign investment and leaves domestic producers without the capital needed to maintain their facilities and international competitiveness. Eventually, the government is forced to resort to external borrowing and may even enlist the assistance of its state-owned mining corporations in this endeavor.

In short, many mining countries appear to have squandered their mineral rents in an effort to avoid the painful economic adjustments required for growth. They have maintained living standards at artificially high levels, even though this has robbed the mineral and other economic sectors of the capital resources needed for development. Sadly, they appear to have used mineral rents the way addicts use drugs—to escape from their problems and to buy a moment's peace at considerable cost to their future welfare.

The perverse use of mineral rents and the poor growth performance of so many of the mining countries Bomsel examines raise two questions. First, are there economic or political forces that keep governments from using their mineral rents more constructively to invest in their future growth? It is perhaps naive to assume governments in very poor countries can withstand pressures to raise present living standards if the resources to do so are available and instead invest in the welfare of future generations. With modern communications and international travel, the example of the developed world is ever present. Moreover, any properly functioning political system must be sensitive to the personal hardship caused by economic dislocation and change. Clearly, policymakers in poor developing countries may have higher political priorities for the use of very limited discretionary funds than investment in the future.

Second, what are the prospects for the future? Is there any possibility the mining countries can extricate themselves from this vicious spiral and use their future mineral rents more constructively? For this to occur, Bomsel suggests, will require fundamental changes in public policy. Although such changes are possible, and may even now be occurring in Chile, Morocco, and certain other developing countries, the poorest of the developing countries, particularly those of sub-Saharan Africa, are desperately short of the human resources required to mount such sweeping changes. The state dominates the economy.

Government employees are responsible for, and have a vested interest in, the current system, including its perverse allocation of mineral rents. Moreover, efforts to change the current distribution of mineral rents will encounter considerable political resistance from other economic groups. The prospects for the poor mineral-exporting countries, though not completely hopeless, are certainly bleak.

Linkages with the Rest of the Economy

Many of the same issues that Bomsel examines are taken up by Philip Daniel. In what is perhaps the most comprehensive essay on the relationship between mineral wealth and economic development in this volume, Daniel addresses the indirect or linkage effects on the economy caused by mining and mineral processing. The central question of his paper is, Has enough been learned about how mineral revenues are transmitted through the economies of lower-income mineral-exporting countries to make appropriate public policy?

Theories of economic development and the historical experience of mineral-exporting countries provide useful insights. Daniel considers three particularly relevant theories of development. The oldest and most orthodox is neoclassical growth theory, where growth occurs by expanding the available supply of factors of production. Capital, foreign exchange, and government revenues are likely to be important constraints on growth, all of which mineral exports can relieve. This theory is the intellectual foundation for the commonsense view described at the beginning of this overview that mineral wealth is a valuable weapon in the struggle to promote economic development.

This orthodox—and optimistic—view was challenged in the 1950s and 1960s by structuralist theories stressing the lack of linkages between mining and the rest of the economy, secular declines in the terms of trade for mineral-exporting countries, and other structural problems. These theories see mineral exploitation holding back economic development and suggest that developing countries should diversify away from mining.

Despite their quite different conclusions regarding the contribution of mineral wealth to economic development, both the neoclassical and structuralist theories anticipate a predetermined outcome over which public policy has little control.

The third and most recent theoretical development emerged from the macroeconomic analysis of open trading economies and focuses on what has become known as the Dutch Disease, an allusion to the experience of the Netherlands and its production of natural gas. This approach predicts that foreign exchange inflows in response to a minerals boom will raise real wages and the price of nontradable goods such as housing and increase the country's real exchange rate. Labor will move to the booming mineral sector and perhaps the nontradable-goods sector, and these sectors will expand. The tradable-goods sector, which includes agriculture and manufacturing, is then crippled in its competition with foreign producers by the movement in real wages and exchange rates and, so, contracts. For this reason, a mineral boom may cause deindustrialization in developed countries, such as the Netherlands or Britain, and a decline in the agricultural sector in developing countries.

These perverse results, however, are not inevitable. They depend on the anticipated upward movement in both the real wage and the exchange rate, which public policy can prevent. In addition, when the assumptions of the Dutch Disease model are relaxed to accommodate the rigidities the structuralist theories highlight, public policy remains of critical importance. Thus, the Dutch Disease model, steering a middle course between the first two theories, concludes that mineral wealth can help or hinder economic growth and that public policy largely determines which of these two possible outcomes prevails.

This conclusion is consistent with the actual experience of non-fuel mineral-exporting countries over the 1965–1985 period. Some of these countries behaved as the Dutch Disease theory anticipates in the absence of offsetting public policies. In particular, the real wage and exchange rate rose when the mineral sector was booming and fell when it was declining. However, the opposite tendencies prevailed when countervailing public policies were implemented.

After concluding that public policies can make a difference—indeed, the difference between a positive and negative outcome—Daniel identifies three particularly important policies. The first addresses what he calls the taxation problem. Here the government's objective is to implement a system of revenue sharing between mineral-producing companies and the state that maximizes the present value of the current and future mineral rents accruing to the government.

This is not as simple as it might sound. Too onerous a tax system destroys the incentives of private companies to explore for and develop new mineral ventures, thus killing the goose that lays the golden egg. Too lenient a tax system reduces the rents flowing to the government and economic development. The nature of the tax system can also affect the efficiency of mineral production and, in turn, the magnitude of the total rents available for companies and the government to divide. Moreover, these difficult issues do not disappear where mining and mineral processing are carried out by state enterprises, because decisions still must be made on how best to divide the rents between the general needs of society and the further development of the mineral sector.

Daniel maintains that enough has been learned about taxation policy to provide some useful advice for policymakers. In particular, he argues for the resource rent tax. With this tax, companies pay regular corporate income taxes until their internal rate of return on invested capital exceeds some prescribed level representing a competitive rate of return on capital. All revenues above that level are taxed at very high rates.

Papua New Guinea and a number of other mineral-producing countries have introduced such tax regimes over the past several decades. In highly cyclical industries such as mining, where profits fluctuate greatly from year to year, the resource rent tax allows the host government to capture a large share of the realized mineral rents without greatly altering the extent, pace, or nature of exploitation that would have prevailed in the absence of the tax. This neutrality is important because it ensures that the rents actually realized and available to divide between companies and the government do not fall far short of their potential.

The second set of important policies Daniel examines deals with maintaining macroeconomic balance. The cyclical instability and other shocks that plague mineral markets cause the rents mineral-exporting countries realize to fluctuate greatly over time. This instability in revenue flows, unless neutralized by public policies, can cause several serious problems at the macroeconomic level. During boom periods, there is pressure for the real wage rate and exchange rate to rise, causing inflation and undermining the competitiveness of domestic agriculture and manufacturing. The surge in government revenues is likely to result in higher public spending. When mineral prices decline and the flow of rents falls, it is difficult to cut back on public programs. A sharp drop in the balance of trade may result in an undesirable surge in foreign debt.

Here again Daniel contends that a great deal has been learned. Public policy should divide the incoming flow of mineral rents into two parts, the first to be invested domestically and the second abroad. The pace of domestic absorption should not exceed the level consistent with the country's capacity for medium-term growth. Attempts to stimulate growth beyond this level simply produce strong inflationary pressures. Moreover, this pace of domestic absorption is only a ceiling. The actual pace should be based on what the fluctuating stream of future rents, conservatively estimated, can maintain over time. This strategy requires that the government add to its assets abroad when mineral prices and rents are high and then draw on these assets when rents are down. For political reasons, such stabilization strategies are hard to pursue, but the difficulties can be reduced by introducing institutional mechanisms that separate mineral revenues from other government income and explicitly call for investing the mineral revenues along the lines described.

The third set of important public policies addresses the absorption problem, or how the mineral rents accruing to the host country are spent. Clearly, if most or all of these returns are used to increase current consumption, rather than invested, mining and mineral processing are not likely to contribute greatly to the country's long-run growth and development. Moreover, even where a large portion of the rents is invested, much depends on how they are invested.

On this issue, Daniel notes that the prevailing wisdom is changing. Historically, most of the mineral-exporting countries within the developing world have relied on the public sector to invest their mineral rents, and state agencies have poured many millions of dollars into roads, ports, and other infrastructure, as well as into new state enterprises in the industrial sector. The generally poor performance of such investments, coupled with worldwide disillusionment with central planning and state ownership, is increasingly calling this strategy into question. Daniel expects that in the future mineral-exporting countries will use much more of their mineral rents to encourage private investments. He urges the low-income mineral-exporting countries to consider fostering more private investment in agriculture. Many of these countries possess a natural comparative advantage in agriculture, and an expansion of this sector would enhance the distribution of income, reduce unemployment and underemployment, and help stem the tide of immigration toward the major urban centers.

The investment programs governments continue to undertake on their own may also change. The experience of the rapidly developing countries, particularly those in the Asia-Pacific region, suggest that more public investment should flow toward human capital, specifically education and health, and less toward physical infrastructure and state enterprises.

Drawing on his analysis of both the theoretical literature and empirical evidence, Daniel concludes:

> There is little a priori inevitability about the outcome of mineral booms and slumps. It is possible to turn a mineral windfall to advantage, but it is also possible to create a development pattern that is worse (in terms of welfare) than that which would have been in place without the minerals. The outcomes depend largely on the things governments do, and thus on the pressures and interests that form and influence governments. . . .
>
> Governments with short time horizons and countries with competing centers of power will continue to operate under the constraints that have contributed to poor performance in the past. Nevertheless, past experience, from African nonfuel mineral exports to 'beneficiaries' of the oil booms, has been sufficiently disappointing and the potential for economic and social development is so great that alternative approaches may now be attractive.

On the basis of both his research and experience as an adviser to a number of developing countries, Daniel ends by offering several rules of thumb for low-income mineral-exporting countries wishing to accelerate their long-run economic development and the contribution their mineral wealth makes to this goal.

Conclusion

The papers in this volume make clear that much is known about the effective use of mineral wealth in promoting economic development. But low-income developing countries for the most part have not been successful in using this asset for economic growth. For many of these countries, mineral resources provide the only significant bridge linking their economies with those of the developed world and offer one of the few hopes they have of attracting the foreign funding and technology they so desperately need if they are to escape their current economic poverty and stagnation.

To begin, the mineral resources of the developing world must be exploited—they contribute nothing to economic development in the ground. Then public policy must be created that maximizes the mineral rents flowing to the host country—a difficult and complex task. Taking too large a share of the rents will discourage exploration and mineral production, reducing the flow of total mineral rents and in turn the host country's share. But taking too little is not desirable either. Finally, the adverse effects of mineral exploitation caused by the instability of commodity markets and other factors need to be minimized, while the beneficial effects, such as jobs for the unemployed and training for the unskilled, are maximized.

Improving the contribution of mineral wealth to economic development is critical not only for the low-income mineral-exporting countries but for developed nations as well, for widespread poverty and the seething frustration it causes threaten the whole world.

Reference

Keynes, John Maynard. 1964. *The General Theory of Employment, Interest, and Money* (New York, Harcourt, Brace and World).

Mining Companies, Economic Nationalism, and Third World Development in the 1990s

THEODORE H. MORAN

The 1970s was a time of broad-based attacks against multinational corporations in the Third World, with the sharpest hostility reserved for investors in the natural resource sector. Something of a turnaround occurred during the 1980s, stimulated by the debt crisis and a perceived need for the capital and management that foreign companies offer the development process. Nations with a long tradition of suspicion toward international companies—from Argentina to India, Mozambique to Yemen, the Andean Pact to the sub-Saharan periphery—began to incorporate foreign firms into their five-year plans.

According to the Association of Political Risk Analysts, there has been a dramatic transformation in political risk climates across the Third World.[1] In the words of *The Economist* (1988), the message is "Come back, multinationals." But will this last? Have we in fact passed a watershed in the treatment of foreign investors, even in the sensitive extractive sector? Although aggregate flows of capital into the mining

---

[1] Annual surveys by the Association of Political Risk Analysts show that concerns about the basic attitudes of host governments toward foreign investment have dropped since 1984 across all regions, including Africa, Latin America, the Middle East, and Asia. For a summary of the more liberal approach to direct investment by region and country, see Aranda (1988).

sector of less developed countries (LDCs) declined in the 1980s, recent years have nevertheless witnessed some of the largest mining investments in Third World history, including the $1.8-billion El Cerrejon coal operation in Colombia and the $1.1-billion La Escondida copper mine in Chile. With exports projected to reach 15 million metric tons (tonnes) in the early 1990s, El Cerrejon could make coal the second-largest (legal) foreign exchange earner for Colombia, behind coffee, greatly contributing to the country's prospects for development. Escondida, an open pit mine in north-central Chile with ore reserves of 660 million tonnes averaging 2.1 percent copper content and with an expected life of 50 years, is slated to become the world's second-largest copper operation in the mid-1990s, with an annual capacity of 320,000 tonnes of contained copper.

As replacement capacity is needed in the next decade, higher ore content and lower extraction costs will tend to favor LDC production sites. If private direct capital continues to flow, mining may once again make a major contribution to Third World development. The prospects for new investment depend, however, on how investors expect to be treated. Will future ventures suffer the same tortured fates as their predecessors, or will they enjoy a new era of stability in their relations with LDC host governments? What awaits international mining companies looking toward the Third World for fresh opportunities?

The Sources of Economic Nationalism

What are the sources of attacks on mining investors in the Third World? The most common hypothesis suggests that economic nationalism springs from sources that are fundamentally nonrational, extra-economic, ideological, or purely emotional. It argues that mining investments do have much to offer LDC governments, but they excite particularly intense feelings about the national patrimony, about subsoil rights, about the God-given (exhaustible) resource base that constitutes the natural heritage of the nation. Regaining control over the extractive sector once it has been developed provides "psychic gratification," a term first used by Johnson (1965). (For a contemporary assessment, see United Nations [1989].) It satisfies what Kindleberger (1969) called "the peasant, the populist, the mercantilist, or the nationalist which each of us harbors in his breast."[2]

To assuage this emotional response, LDC leaders and their populations must be convinced that the benefits of natural resource investment can be divided in ways that allow both sides to prosper, even in settings where there is ingrained hostility to capitalism, as in Angola. Mutual benefits, exemplary corporate conduct, and careful persuasion, the hypothesis suggests, are key to overcoming the nonrational appeal of economic nationalism. The passage of time may be on the side of moderation. As Third World governments gain a greater understanding of international mineral markets and acquire greater ability to monitor and even control (or seek outside help and advice in monitoring and controlling) how mining companies operate, the lingering feelings of resentment, suspicion, and tension ought to dissipate.

If this nationalism-as-psychic-gratification hypothesis alone were to explain the dynamics of investor–host relations, it might be plausible to conclude that the treatment mining investors got up until the debt crisis of 1982 was in fact a stage that developing countries might pass through and move beyond. Rendering this hopeful appraisal more problematic, however, is a second cluster of hypotheses about the relations between foreign mining investors and host authorities, growing out of the idea of the "obsolescing bargain" (Vernon, 1971; Moran, 1974; Sklar, 1975; Smith and Wells, 1975; Tugwell, 1975). Here the argument asserts that economic nationalism is in some sense a manifestation of quasi-rational self-interest.[3] In the obsolescing-bargain model, the demand for fundamental renegotiation of mining investment agreements results from the dissipation of risk and uncertainty after projects requiring large sunk capital prove successful. Just as foreign investors cannot avoid demanding generous terms when the initial risk and uncertainty are high, host authorities cannot avoid demanding that the terms be revised once the risk and uncertainty are gone. Although there may be popular emotion and even hysteria accompanying the renegotiations, the adjustment

---

[2] In his own analysis, Kindleberger goes beyond nationalism as pure emotion to develop the idea of bilateral monopoly bargaining.

[3] *Rational* means that there is an underlying logic of self-interest to the process even though the renegotiation of investment agreements may be pushed along by highly emotional popular rhetoric.

process itself has an underlying logic and rationality that render it inevitable.

If the obsolescing-bargain model is the correct model to apply to future business–government relations in the LDCs, there are ominous implications for the role that mining may play in Third World development. The renegotiation of mining agreements tends not to be merely a self-adjusting mechanism, with early extravagant benefits to investors being offset later by larger tax shares for hosts. Rather, if the 1970s is to be a guide to the 1990s, this process, Mikesell (1975) argues, tends to overshoot (even when there is no overt nationalization), undercompensating investors for the capital tied up in the exploration and development phases. After examining in detail the internal rate of return on investment of the Toquepala mine in southern Peru and the Bougainville mine in Papua New Guinea, Mikesell (1975, p. xx) concludes that had the "changes in the investment climate been anticipated by the foreign investors, it is doubtful whether either of these mines would have been constructed."

In short, the obsolescing-bargain model represents a case of market failure for the international mining industry. It skews mineral investments away from locales in the Third World where the geological endowment and production costs would otherwise constitute a comparative advantage. Contrary to the hopes of private investment enthusiasts, the problem cannot be overcome simply by the promise of a "good investment climate" during times (such as today) when host authorities want to attract new mining companies.

But is the future of foreign investment in the mineral industries of Latin America, Africa, Asia, and the Middle East so dismal? Does more recent experience point to such a difficult operating environment for investors in the extractive sector? Because of low prices for industrial commodities, the period from the mid-1970s to the mid-1980s does not offer much evidence about how new mining agreements in the Third World may evolve. But a comparative perspective from the evolution of business–government relations in sectors other than mining may help.

The Structural Vulnerability of LDC Mining Investments

In the 1980s the study of foreign investor and host government relations broadened considerably in sectors other than natural resources. The analysis of industries that range from agribusiness to chemicals, automobiles to pharmaceuticals, and computers to consumer electronics provides a database that is not skewed by the preoccupation with subsoil rights, exhaustible resources, and national patrimony that may make mining (and petroleum) a special case.

What insights do the nonextractive cases offer for forecasting the outlook for mineral projects in the 1990s? For one, they suggest that there are many kinds of investments that may well be relatively immune to the squeeze of economic nationalism in the coming decade. Unfortunately, mining projects are not likely to be among them. There is a "structural vulnerability" associated with mining activities that makes foreign companies particularly liable to the renegotiation and tightening of investment agreements.

An important factor across industries that determines whether individual projects are likely to be vulnerable to economic nationalism is the size of the sunk investment. Companies that have relatively small fixed commitments ($50 million or less) can credibly threaten to withdraw if host authorities push them too far. IBM left India, General Motors left Peru, and Dow left Korea when each had relatively small operations being squeezed (Bergsten, Horst, and Moran, 1978; Schwendiman, 1984; Grieco, 1985). Companies that require large fixed investments, in contrast, do not enjoy such flexibility. Instead, they are trapped once their activities are in place. International companies in nonextractive sectors that require large fixed investments (world-scale operations in chemicals, petrochemicals, fertilizers, and automobiles) have experienced the same "hostage effect" as mining and petroleum (Teeple, 1983; Schwendiman, 1984). In addition, the size of the investment may make it more visible and politically sensitive.

An examination of the role technology plays in determining the evolution of business–government relations further strengthens the concept of structural vulnerability.[4] According to Kobrin (1980), the maturity of the technology used in the project has a statistically signifi-

---

[4]The issue here is the role technology may play in strengthening or weakening the foreign investor relative to governmental authorities, not the more familiar debate about whether foreign investors use "appropriate technology" for Third World development.

cant impact in predicting the likelihood of nationalization.[5] Bradley (1977) finds that midrange technology firms (as measured by the level of research and development expenditures) have been liable to successful attack by host country nationalists. Lecraw (1984) concludes that technological leadership was a significant factor in determining bargaining outcomes in 153 subsidiaries from six industries in Thailand, Malaysia, Singapore, Indonesia, and the Philippines. This author's work (Moran, 1985) highlights the dichotomy between dynamic and stable technology: The former shields the investor from nationalistic pressures, but the latter does not.

Overall, the evidence suggests that incorporation of process technology in any given foreign project that is not highly sophisticated or changing rapidly, and that therefore can be replicated easily with no more than acceptable losses in efficiency, renders the foreign investment project open to successful assaults by local authorities. Transposing these findings to future mining contracts is complicated because there is a good deal of innovation in mining technology, especially in sensor and computer applications related to the exploration phase. But much of the control over this sophisticated and dynamic aspect of mining operations is expended, so to speak, before the funds are committed and therefore cannot be used to offset the vulnerability of the on-line project.

Besides the size of the fixed investment and the stability of the technology, a third variable in determining the course of business–government relations is the extent of product-differentiation, brand identification, or consumer loyalty created by advertising. Fagre and Wells (1982) showed that international investors who spend more than 7 percent of total sales revenues on advertising are particularly immune to nationalistic demands. Lecraw (1984) likewise discovered a statistically significant effect from advertising intensity. These findings do not provide much solace to potential mineral investors. Although product differentiation may prove helpful to some firms in the Third World, it has never been much of a factor in mining.

---

[5]A weakness found in most of the comparative statistical analyses is that they take ownership as the dependent variable, that is, as the principal measure of multinational corporation or host government bargaining strength (see Kobrin [1987]).

A final variable affecting the evolution of business–government relations is the extent of competition in the industry. Knickerbocker (1973) was the first to discover a clearly identifiable "burst" phenomenon in which companies follow closely on the heels of the first investor in an industry; this follow-the-leader effect has also been demonstrated in more recent studies (Yu and Ito, 1988).

The creation of competition (or potential competition) among investors provides more alternatives for host authorities to choose from, thereby strengthening their hand, much as the dialectic between independents and majors in the oil industry has done.[6] The presence of a maverick in the automobile, tire, and electronics industries has increased the ability of host negotiators to impose new demands (Bennett and Sharpe, 1985; West, 1985; Doner, 1988).[7] Conversely, lack of competition helps the stability of the foreign investor. Fagre and Wells (1982), for example, found that the strongest bargaining success in 16 of 18 Latin American countries occurred when a multinational investor found itself to be the sole foreign company present in the industry.

Once again, natural resource investment is not favored. As concentration ratios in mining industries decline, one should expect added pressure on the stability of mining agreements.

Overall, studies of the evolution of business–government relations across multiple sectors in the Third World indicate that there is a structural vulnerability that is not associated with primordial feelings about subsoil resources.[8] Furthermore, from a comparative perspective, mining industries do not look promising, and would not even if the psychic need to directly control national patrimony were to lessen.

What are the consequences for Third World development? Should we expect the future of business–government relations in mining to reflect the past, meaning market failure for new investments? Or

---

[6]For evidence from both the petroleum and mineral sectors, see Moran (1979).

[7]Japanese firms are not immune from competing against each other to the advantage of host authorities. Doner (1988) finds that Japanese firms also manifest strong rivalry and competition within oligopolies.

[8]Of interest to political risk analysts is the fact that structural vulnerability and dynamics of the obsolescing bargain appear to operate independently of ideology, orientation, or the military–civilian composition of government.

are companies developing new strategies and techniques to mitigate their own vulnerability?

Strategies to Offset Investor Vulnerability

One might suppose that the sudden interest in political risk as an element in international corporate strategy that followed the fall of the Shah of Iran would have produced sound advice for future investors in the extractive sector. Unfortunately, this is not the case.

The literature on how international investors should deal with political risk makes eminently good sense. Springing from the community of financial analysts in corporations, consulting firms, and business schools, it advises investors to compensate themselves for the added risk they face by raising the hurdle rate the project must pass before the corporation approves the investment decision (Brewer, 1981; Teeple, 1983; Lessard, 1985).

This conventional approach fits well with the avalanche of advice being offered by the International Monetary Fund, the World Bank, and authorities in Washington, D.C. It focuses on the need to establish a good investment climate as a prerequisite to attracting multinational investors. And it seems to suggest that if entry conditions can be made sufficiently favorable, foreign-sponsored projects should move forward.

Such advice is of very little practical use to multinational companies that face situations of structural vulnerability in their industries. For one thing, the difficulty with mining ventures does not arise from stingy entry conditions; rather, it comes from the likelihood that the terms of the contract will be challenged by economic nationalists once the project is successfully on-line. Further, the conventional financially oriented approach to political risk suggests that good projects that cannot be successfully front-loaded (this would include all but a few oil plays in the petroleum industry when prices are high) should not be approved.[9] This stance simply affirms the essence of market failure for the international mining industry.

---

[9]In the highly risky Angolan oil ventures, for example, the international companies required, and got, a payback period of less than three years.

International mining companies are left, therefore, with two equally unappealing options: (I) to search evermore for enthusiastic and compliant hosts who probably will not be able to make good on their promises, or (2) to give up on projects that are commercially attractive.

Strategies to Manage Political Risk

What is needed are strategies that allow projects whose basic cost characteristics under foreign developers are favorable to go ahead despite the fact that they face an unavoidable exposure to economic nationalism after coming on-line. Here the literature that may be of use to international mining strategists is sparse and comes from industries whose experiences cannot always be generalized. Still, a comparative survey across other business sectors might suggest effective approaches for dealing with political risk for structurally vulnerable industries, and in particular, mining.

One such approach is to control access to downstream markets to thwart the attempts of economic nationalists to move in on upstream operations. In this context, tropical agribusiness firms have deliberately diversified sources of supply at the plantation stage while retaining an advantage in rapid and efficient (refrigerated) transport to retail buyers in the consuming countries (Litvak and Maule, 1977). Electronics manufacturers have situated labor-intensive stages in the Third World while maintaining research and development facilities and specially designed test facilities at home.[10]

This strategy provides the international company with a monopsonistic advantage over upstream producers (reinforced in vertically integrated industries by favorable transfer pricing possibilities). It simultaneously raises the price for successful economic nationalism at the production end, requiring would-be nationalists to overcome the hurdles in other stages with higher barriers to entry in order to exploit what pressures they can apply at the production end. Overall, across industries, the degree of subsidiary integration within international

---

[10]These findings are also supported by interviews the author held with agribusiness and electronics firms from 1984 to 1988.

companies has been shown to be a statistically significant variable in reducing the bargaining effectiveness of host authorities (Fagre and Wells, 1982; Lecraw, 1984).

How helpful is the idea of recapturing oligopolistic control downstream, outside the reach of economic nationalists at the production stage? The option is real, but highly limited. In the aluminum industry, the stage with the highest barriers to entry (capital costs and access to energy) has traditionally been the smelting stage rather than the bauxite mining stage. To preserve their leverage over producing countries, the international companies have been deliberately slow in building smelter capacity within the same jurisdictions as bauxite operations despite the offer of cheap natural gas or government-subsidized hydroelectric power to induce smelter investments (Moomy, 1984). In the oil industry, levels of concentration in the tanker, refiner, and retail stages have tended to be more diffuse than at the production stage, making the effort to influence upstream decisions from the vantage of a choke point downstream difficult (although some international firms, notably Shell, appear willing to try [Moran, 1987]). In many extractive industries, concentration ratios at the smelting and refining stages are lower than at the mining stage. Thus, denial of access to downstream processing is not likely to be a major strategic option for future business planners in the international mining industry.

A second approach to attempting to limit host country nationalism has been to offer joint ownership of a project. Behind this option is a promising rationale: Shared ownership may defuse charges of control by outsiders, gain the foreign investor an ally, and remove the mantle of mystery and suspicion that otherwise surrounds multinational corporate operations. And it almost always proves attractive to host negotiators.

Comparative data only partly support such a strategy. Kobrin (1980) finds, for example, that although joint ventures have indeed been nationalized less often than wholly owned subsidiaries in the Third World, there appears to be a clearly identifiable "first bite" phenomenon in which joint ventures with a host government have whetted the government's appetite for (and perhaps provided managerial expertise to facilitate) some kind of subsequent takeover, intervention, or contract renegotiation. Only joint ventures with local private partners actually lowered the investor's vulnerability to such hostile acts. Joint ownership with a state mining company, therefore, is far from a panacea for the threats of economic nationalism.

A third approach to managing a situation in which vulnerability to economic nationalism is expected to increase is to sequence the way in which the investment is carried out. This tactic divides the project into phases, so as to have a major new increment to offer when host authorities demand more. There is evidence that agrichemical industries, for example, have undertaken fertilizer investments in carefully spaced blocks solely to maintain a card to play in the face of nationalistic pressures (Bergsten, Horst, and Moran, 1978). Chemical and petrochemical strategists have overridden engineers (who wanted complete plants) and economists (who wanted lowest unit cost) as a tactic to keep seemingly insatiable host authorities at bay. Manufacturing firms have consented to move from assembly operations to local production and even exports as part of an ongoing process of defraying demands for higher taxes or greater local ownership (Bergsten, Horst, and Moran, 1978; Bennett and Sharpe, 1985). Ford and IBM, for example, have maintained their status as exceptions to Mexico's statutory ceiling of 49 percent foreign ownership by "giving in" to the country's desires for higher value added and a greater contribution to the Mexican balance of payments (Business Week, 1984; New York Times, 1985; Kobrin, 1987).

In the mining industry the idea of sequencing corporate commitments offers interesting potential. In Peru, Asarco agreed to develop a new mine, Cuajone, with funds spared from high taxes on its existing mine, Toquepala (Mikesell, 1975). Other mining investors have offered more value added to their operations by installing processing capacity near Third World centers of production; how much of this might have resulted from attempts to reduce nationalistic pressures on the production stage, as opposed to conventional reasons of comparative advantage, is not clear. The use of tax-sheltered proceeds from an existing mine to finance a new one or the prospect of adding smelters or refineries over the life of a given operation does suggest that a strategy of deliberately sequencing the investment process as a way of meeting new nationalistic pressures might be an option for the international mining community.

In general, however, it is clear that these three approaches offer rather meager prospects to multinational mining companies that want to develop commercially promising projects in settings where pressures to renegotiate the original agreements are all but inevitable. In copper, nickel, tin, and coal mining, opportunities for controlling access to downstream markets via a clamp on processing are not readily available. Joint ventures with host authorities may assuage nationalistic feelings for a while but might become the first step toward an eventual takeover. Sequencing the investment process may postpone the squeeze of nationalism, but it requires ever-greater commitments of corporate resources with no guarantee of permanency.

These findings must also be disappointing to Third World host authorities who want to provide adequate conditions for foreign mining investment but who know that their own successors are unlikely to honor the terms necessary to get the corporations to enter the country in the first place. They find themselves in the position of negotiators who, in the words of Schelling (1963), can only "cross their hearts" to show their promises are sincere. The dilemma is that they cannot in fact make their commitments credible. This is the weakest of all positions for a negotiator to be in.

Use of Project Finance to Offset Political Risk

A final way to handle economic nationalism for both foreign investors and host country negotiators may be found within the extractive sector itself, in the way petroleum companies (and some mining companies) have been using project finance to construct an alliance of creditors who can put clout behind assurances of stability for the original investment agreement. The idea is to build a structure of lenders (including the commercial banks and export-import banks of a number of industrial countries) whose repayment prospects will be altered by a dramatic change in the status of the foreign operator. Given the dependence of most LDCs on unimpeded access to international financial markets for trade credits, supplier credits, and the rollover of their longer-term debt, the potential disapproval of major segments of the financial community should have a sobering effect on those ministers and cabinets who succeed the signers of the original investment agreement (for a general outline of this approach, see Moran [1985] and Waelde [1988]).

This approach is not completely new to the international mining industry. In 1975 Asarco formed a syndicate of 29 banks from all the major industrial countries to reduce its capital exposure and spread risk in developing the new Cuajone mine in Peru (the company changed its debt-to-equity from the conventional 2:1 ratio to 6:1). In addition, with repayment of the loans backed by timely delivery of copper from Cuajone and not by the full faith and credit of Asarco, Asarco's management calculated that the project financing structure would help reinforce its position in Peru as manager of Cuajone (Mikesell, 1975).[11] Moreover, the fundamental investment agreement, completed in 1975, required the foreign exchange proceeds from long-term sales contracts for Cuajone's output to be deposited in an account in the lead New York bank that syndicated Cuajone's loan. Although the account remained in the name of the Central Reserve Bank of Peru, the project's investors were authorized to withdraw payments in the amount of profits, depreciation, and amortization each month. The text of the irrevocable order to the New York bank, subject to the laws of the state of New York and enforceable by attachment in its courts, was incorporated in the investment agreement.

Broken Hill Proprietary (BHP) of Australia used many of the same project-financing procedures in opening the Ok Tedi mine in Papua New Guinea to gold/copper production in 1984.[12] In return for (minority) equity capital, BHP wrote long-term supply contracts with major German refineries, on the basis of which it then secured large project loans ($100 million) from the German Export-Import Bank (Kreditanstalt fuer Wiederaufbau) and from the development bank of the European Economic Community (European Investment Bank). The ongoing presence of BHP as owner–manager of Ok Tedi, unencumbered by burdensome regulations on the part of the newly independent government of Papua New Guinea, was the "guarantee" of uninterrupted delivery. Should any fundamental change of management or operating environment interrupt supply, there would be counter-

[11] This point was also made in interviews the author held with the chairman and the vice chairman of Asarco from 1984 to 1986.
[12] Broken Hill Proprietary was the lead investor in a consortium that included Amoco Minerals and a group of West German refineries (for details, see McGill [1983]).

pressure from major actors in Papua New Guinea's principal metals market. In addition, BHP nominated Citicorp-Australia as the lead bank for a group of 15 commercial lenders, with a trust account to which all future sales proceeds were assigned for distribution by Citicorp.

In one of the largest recent mining agreements—the $1.8-billion El Cerrejon coal-export project in Colombia, signed in 1982 with exports commencing in 1986—Exxon formed a joint venture with Carbocol, the Colombian state oil company (Kline, 1988). Holding large cash reserves after the oil price rise of 1979–1980, Exxon decided to finance its share of the operation entirely in-house. Nevertheless, Exxon may have obtained some multilateral protection by securing its sources of equipment financing from diverse sources, including Canada and Great Britain, and not just from the United States. The deterrent value of the supplier credits (7 years) is probably less, however, than were the diversified commercial loans in Cuajone and Ok Tedi (10–12 years).

Such an approach at best helps ensure the status of the foreign investor as owner and operator of the mine for the duration of the financial package. What about seeking protection against major changes in the fiscal terms of the investment agreement, the tax rate, depreciation and amortization schedules, and access to foreign exchange?

Here the petroleum industry has gone a step beyond the mining companies (Perera, 1987). Several oil companies, including Occidental, Chevron, Phillips, and Elf Aquitaine, have incorporated World Bank loans within the syndicate of lenders to surround the private financial contributors with World Bank cross-default clauses; that is, their projects cannot default on any of the loans without also being in default to the World Bank. In addition, they have included in the remedies clauses of the World Bank loans commitments from the host governments not to alter the terms of the investment agreements (or the laws surrounding them, including the tax laws) in such a way "as to, in the opinion of the Bank, materially and adversely affect the project." If the host violates this pledge, the World Bank can stop disbursement on the loan, demand immediate repayment of the balance of the loan, and ultimately call in the entire portfolio of outstanding loans to the country in question. Finally, they have encouraged and taken advantage of a new World Bank cofinancing initiative ("cofinancing B loans") in which the World Bank's loans are concentrated in and guarantee the later maturities in the financial package. The goal is to extend the involvement of the World Bank past the end of the payback period and well into the time when the risk of the obsolescing bargain is growing.

A further possibility, not yet contained in any investment agreement, is to use "sponsor guarantees" in which the World Bank lends directly to a foreign-owned project, with a "political force majeure" clause allowing the investor-controlled project to suspend repayment to the World Bank if the host violates the commitments made in the initial investment contract. If implemented, this would allow the parent company to trigger a default to the World Bank that the latter would then pursue with the host (Walser, 1983).

In all these petroleum cases, the objective is to use the clout of the World Bank to counter attempts to make drastic changes in the terms of the original investment agreement. The international mining industry may be heading in the same direction. In the $1.1-billion Escondida copper project in Chile, signed in 1987 with commercial production beginning in 1991, the principal investor, Broken Hill Proprietary of Australia, chose the International Finance Corporation (IFC), an agency of the World Bank, to help with syndication of the $680-million loan package and to take a 2.5 percent equity position itself.[13] According to BHP, this spreads an IFC umbrella of political protection over the equity holders, including BHP itself (57.5 percent), Rio Tinto-Zinc (30 percent), and a Japanese consortium headed by Mitsubishi (10 percent). A change in the tax rate (fixed at 49.5 percent for 20 years in the investment agreement with the government of Chile) would directly affect the IFC. An act of expropriation would be considered a material breach of the broader agreement that covers all the IFC's activities in Chile, which proscribes any form of "seizure" by executive or legislative action. Although neither a tax change nor a nationalization would legally require the World Bank to take action with regard to its extant portfolio of loans to Chile, the Bank, according to IFC officials, could not help but take note of such an abrogation of contract as it considered new Chil-

[13]See *The Broken Hill Proprietary Company Limited*, Form 20-F, submitted to the U.S. Securities and Exchange Commission, November 30, 1988; and "Escondida Project in Chile Given Go-Ahead," BHP-Utah, San Francisco, July 25, 1988.

ean loan applications, since "the Board of Directors of the IFC is essentially the same as the Board of Directors of the World Bank."[14]

Further support for the stability of BHP's investment agreement may come from the buyers and lenders. Three-quarters of the output to the year 2002 has been committed under 12-year sales contracts to smelters in Japan, Germany, and Finland, giving them a substantial interest in a steady flow of output. At the same time, the long-term contracts have been used to generate import financing from the Export-Import Bank of Japan ($350 million), the Kreditanstalt fuer Wiederaufbau of Germany ($140 million), and the Kansallis-Osake-Pankki of Finland ($47 million) for the full 12-year period. An interruption in production, in the words of BHP, "would have the Export-Import Bank of Japan banging on the door of the Chilean government" (personal communication with BHP executives, January 1989).

Finally, all the financing was done on a nonrecourse basis, with the Chilean subsidiary (Minera Escondida Limitada), not the BHP parent, as borrower. The Industrial Bank of Japan was appointed as trustee for the lenders, with proceeds of all sales paid to an account it holds outside Chile. It disburses the resulting monies to all parties according to the terms of the original investment agreement throughout the 20-year duration of the agreement.

Conclusion

What can be concluded from this review about the likely future for international mining investments in the Third World? For one thing, there probably will be an ongoing debate among corporate strategists about the pros and cons of joint ventures and ongoing attempts at sequencing mining and mineral processing projects so as to placate the demands of nationalists. But these techniques are unlikely by themselves to compensate for the structural vulnerability that mining investors will most likely continue to experience. To counter this fundamental

---

[14]Interviews with IFC officials, December 1988. Adding to its clout, the IFC itself has a portfolio worth $200 million outstanding in Chile, representing a total of $2 billion in direct investments. On the other hand, a host government may consider the option of buying out the IFC share prior to demanding a renegotiation of the original agreement.

structural vulnerability, there probably will be further experimentation in the extractive sector with financial structures aimed at deterring major changes in original investment agreements. The objective will be to bind the hands of the political successors who follow those who sign the initial documents.

Is progress in this direction in the interest of the Third World? After more than a decade of movement toward greater autonomy for Third World countries, is this not a retrograde prospect? To be sure, the strategies for mining investment discussed here constitute a deliberate step away from unfettered freedom of action for LDC host authorities. Such strategies will push large-scale mining agreements in the direction of solemn treaty obligations that carry grave consequences if broken. At the same time, however, they will strengthen the ability of Third World negotiators to make their commitments credible, greatly enhancing their negotiating leverage. These techniques to manage political risk should not totally prevent the readjustment of mining contracts as risk and uncertainty dissipate or as mineral prices fluctuate; instead, they may simply help push legitimate disputes in a more productive (less zero-sum) direction. They should allow promising but risky projects to go forward that otherwise would be left on the drawing boards, enhancing greatly the contribution that mining can ultimately make to Third World development.

Acknowledgment

I wish to thank Richard L. Gordon, Thomas Waelde, and Louis T. Wells, Jr., for helpful comments on earlier versions of this paper.

References

Aranda, Victoria. 1988. "National Policies on FDI in the 1980s," The CTC Reporter vol. 26 (Autumn) pp. 34–37.

Bennett, Douglas C., and Kenneth E. Sharpe. 1985. "The World Automobile Industry and Its Implications," in Richard Newfarmer, ed., Profits, Progress, and Poverty: Case Studies of International Industries in Latin America (Notre Dame, Ind., Notre Dame University Press).

Bergsten, C. Fred, Thomas Horst, and Theodore H. Moran. 1978. American Multinationals and American Interests (Washington, D.C., Brookings Institution).

Bradley, David G. 1977. "Managing Against Expropriation," *Harvard Business Review* vol. 55, no. 4 (July–August) pp. 75–83.

Brewer, Thomas L. 1981. "Political Risk Assessment for Foreign Direct Investment Decisions: Better Methods for Better Results," *The Columbia Journal of World Business* vol. 16, no. 1, pp. 5–12.

Business Week. 1984. "Ford's Better Idea South of the Border," January 9.

Doner, Richard. 1988. "Weak State–Strong Country? The Thai Automobile Case," *Third World Quarterly* vol. 10, no. 4 (October) pp. 1542–1564.

The Economist. 1988. "Come Back Multinationals" (review of the 1988 Report of the United Nations Centre on Transnational Corporations), November 26, p. 73.

Fagre, Nathan, and Louis T. Wells, Jr. 1982. "Bargaining Power of Multinationals and Host Governments," *Journal of International Business Studies* vol. 13, no. 2 (Fall) pp. 9–23.

Grieco, Joseph M. 1985. "Between Dependency and Autonomy: India's Experience with the International Computer Industry," in Theodore H. Moran, ed., *Multinational Corporations: The Political Economy of Foreign Direct Investment* (Lexington, Mass., D. C. Heath).

Johnson, Harry. 1965. "An Economic Theory of Protectionism," *Journal of Political Economy* vol. LXXII (June) pp. 256–283.

Kindleberger, Charles P. 1969. *Six Lectures on Direct Investment* (New Haven, Conn., Yale University Press).

Kline, Harvey F. 1988. *The Coal of Cerrejon: Dependent Bargaining and Colombian Policy Making* (University Park, Pennsylvania State University Press).

Knickerbocker, Frederick T. 1973. *Oligopolistic Reaction and Multinational Enterprise* (Cambridge, Mass., Harvard University Press).

Kobrin, Stephen J. 1980. "Foreign Enterprise and Forced Divestment in the LDCs," *International Organization* vol. 34, no. 1 (Winter) pp. 65–88.

——. 1987. "Testing the Bargaining Hypothesis in the Manufacturing Sector in Developing Countries," *International Organization* vol. 41, no. 4 (Autumn) pp. 609–638.

Lecraw, Donald J. 1984. "Bargaining Power, Ownership, and Profitability of Transnational Corporations in Developing Countries," *Journal of International Business Studies* vol. 15, no. 1 (Spring–Summer) pp. 27–43.

Lessard, Donald R. 1985. *International Financial Management* (New York, Wiley).

Litvak, I. I., and C. J. Maule. 1977. *Transnational Corporations in The Banana Industry: With Special Reference to Central America and Panama* (New York, United Nations Economic Commission for Latin America).

McGill, Stuart. 1983. "Project Financing Applied to the Ok Tedi Mine: A Government Perspective," *Natural Resources Forum* vol. 7, no. 2, pp. 115–129.

Mikesell, Raymond F. 1975. *Foreign Investment in Copper Mining: Case Studies of Mines in Peru and Papua New Guinea* (Baltimore, Md., Johns Hopkins University Press for Resources for the Future).

Moomy, Ruthann C. 1984. "The Location of Minerals Processing." Working Paper, International Institute for Applied Systems Analysis, Laxenburg, Austria (April).

Moran, Theodore H. 1974. *Multinational Corporations and the Politics of Dependence: Copper in Chile* (Princeton, N.J., Princeton University Press).

———. 1979. "The International Political Economy of Cuban Nickel Development," in Cole Blasier and Carmelo Mesa-Lago, eds., *Cuba in the World* (Pittsburgh, Pa., University of Pittsburgh Press).

———. 1985. "International Political Risk Assessment, Corporate Planning, and Strategies to Offset Political Risk," in Theodore H. Moran, ed., *Multinational Corporations: The Political Economy of Foreign Direct Investment* (Lexington, Mass.: D. C. Heath).

———. 1987. "Managing an Oligopoly of Would-be Sovereigns: The Dynamics of Joint Control and Self-Control in the International Oil Industry, Past, Present, and Future," *International Organization* vol. 41, no. 4 (Autumn) pp. 575–607.

New York Times. 1985. "IBM Concessions to Mexico," July 25.

Perera, Srilal. 1987. "Techniques of Protecting Foreign Investments Against Political Risk" (Ph.D. dissertation, Georgetown University, Washington, D.C.).

Schelling, Thomas C. 1963. The Strategy of Conflict (New York, Oxford University Press).

Schwendiman, John S. 1984. "Managing Environmental Risk: Cases and Lessons for Corporate Strategy," in Fariborz Ghadar and Theodore H. Moran, eds., *International Political Risk Management: New Dimensions* (Washington D.C., Landegger Program in International Business Diplomacy, Georgetown University).

Sklar, Richard L. 1975. *Corporate Power in an African State: The Political Impact of Multinational Mining Companies in Zambia* (Berkeley, University of California Press).

Smith, David N., and Louis T. Wells, Jr. 1975. *Negotiating Third World Mineral Agreements: Promises as Prologue* (Cambridge, Mass., Ballinger).

Teeple, William. 1983. "Integrating Political Risk Considerations into the Capital Budgeting Process," in Fariborz Ghadar, Stephen J. Kobrin, and Theodore H. Moran, eds., *Managing International Political Risk: Strategies and Techniques* (Washington, D.C., Landegger Program in International Business Diplomacy, Georgetown University).

Tugwell, Franklin. 1975. *The Politics of Oil in Venezuela* (Stanford, Calif., Stanford University Press).

United Nations. 1989. *Permanent Sovereignty over National Resources.* Report of the Secretary General, Economic and Social Council, Committee on National Resources, Eleventh Session; 27 March–5 April (E/C. 7/1989/5).

Vernon, Raymond. 1971. *Sovereignty at Bay: The Multinational Spread of U.S. Enterprises* (New York, Basic Books).

Waelde, Thomas W. 1988. "Third World Mineral Investment Policies in the Late 1980s: From Restriction Back to Business," *Mineral Processing and Extractive Metallurgy Review* vol. 3, pp. 121–182.

Walser, Christian. 1983. "Multilateral Institutions and Political Risk: Deterrence, Co-Financing and Compensation," in Fariborz Ghadar, Stephen J. Korbin, and Theodore H. Moran, eds., *Managing International Political Risk: Strategies and Techniques* (Washington, D.C., Landegger Program in International Business Diplomacy, Georgetown University).

West, Peter J. 1985. "International Expansion and Concentration of the Tire Industry and Implications for Latin American," in Richard Newfarmer, ed., *Profits, Progress, and Poverty: Case Studies of International Industries in Latin America* (Notre Dame, Ind.: Notre Dame University Press).

Yu, Chwo-Ming J., and Kiyohito Ito. 1988. "Oligopolistic Reaction and Foreign Direct Investment: The Case of the U.S. Tire and Textiles Industries," *Journal of International Business Studies* vol. 19, no. 3 (Fall) pp. 449–460.

Economic Development and the Timing of Mineral Exploitation

MARIAN RADETZKI

A number of developing countries are richly endowed with mineral resources. After the mines have been built and processing plants established, the mineral sector can play a crucial role in economic progress. In many developing countries, mineral exploitation is a major source of export revenue, wages, and government income.

There is considerable justification for regarding the mineral endowment of a country as a wasting resource. What is extracted today will not be available for use tomorrow. When mineral resources are viewed this way, the timing of their exploitation becomes an important issue.

In general, early exploitation of a finite mineral endowment is preferable unless there is an unequivocally rising price trend. With positive interest rates, the net present value of today's income is greater than the net present value of the same income earned tomorrow. Because economic progress is desirable, and because the income generated by mineral exploitation should speed such progress, there is a net social benefit in exploiting the mineral resources as soon as possible. Nonetheless, various arguments have been put forth in support of policies to defer the development and exploitation of mineral deposits with a profit potential. These arguments either refute the rationality and justice of net present value maximization or assert that early exploitation will not result in maximum present value.

The view that minerals should be left in the ground for the use of future generations often attracts political attention. Its main tenet is that the continued well-being of humankind is crucially dependent on the availability of unextracted exhaustible resources and that the discounting used in present value calculations gives insufficient weight to human welfare in the distant future (Kneese, 1989). Advisers to poor, underdeveloped nations have occasionally argued in favor of deferral, pending development of national institutions and human capital which can assure that carelessness and waste in the extraction and use of the mineral rent, or the foreign appropriation of it, is avoided. When the potential production units are large and indivisible, deferral has sometimes been recommended to avoid a possible social loss from the downward shift in price. Finally, creation of new capacity has occasionally been deferred as producers try to raise prices by cutting supply.

The Hotelling rule, according to which the optimal drawdown of the resource in the ground raises its price at the rate of interest, has supported deferral because it implies that the present value of income earned by mineral exploitation is the same regardless of when exploitation occurs. Thus, nothing is lost by waiting.

In numerous cases, some of which are discussed in this paper, the arguments in favor of deferral have won sufficient political support to result in policies aimed at arresting or slowing mineral exploitation. This paper scrutinizes the validity of these arguments, beginning with the Hotelling rule. It should be emphasized that the arguments in favor of deferral concern projects with a profit potential. No justification is needed for the deferral of projects deemed to be uneconomical or unprofitable at the price levels expected to prevail when the projects become operative.

The Fallacies of the Hotelling Rule

In its original formulation, exhaustible-resource theory claimed, with impeccable logic, that in an environment of profit-maximizing firms using unchanging technology to exploit a natural resource from a known, finite, and uniform stock, the value of a unit of the unexploited resource (the mineral rent or royalty) rises at the rate of interest (Hotelling, 1931). The relationship is commonly referred to as the Hotelling rule. However, the rule applies only to very special circumstances seldom encountered in the real world.

First, mineral resources in the real world vary widely in quality. Each quality category has its own royalty level in the exhaustible-resource model, and all these royalties increase over time at the rate of interest. Economic logic asserts that the best quality with the highest royalty will be exploited first, since this assures the maximization of net present value of the mineral activities. As the best quality is depleted, there is a downward adjustment in the average level of the rising royalties. Still, the royalty on a deposit of a given quality rises over time at the rate of interest, so the basic tenet for deferral of exploitation continues to hold.

Discovery of new deposits through exploration further complicates the Hotelling rule. Because discovery adds to the stock of unexploited resource, the unit cost of discovery at the margin must correspond to the unit value of that resource, that is, to the royalty. If the royalty rises at the rate of interest, the correlation will be maintained only if the marginal cost of discovery increases in equal measure. This holds in the real world only in exceptional cases. Unanticipated massive discoveries unrelated to the exploration expenditure are known to occur from time to time. The unit cost of discovery then falls, and, in consequence, so does the royalty. Hence, the Hotelling rule breaks down, and the entire price path must be reconsidered. With unanticipated discovery, even the assertion that the royalty related to a deposit of a given quality rises at the rate of interest ceases to hold.

Further questions about the validity of the Hotelling rule arise when technological progress is introduced. Consider first the technology of exploration. It is not possible to assert that when improvements in exploration technology are introduced, the unit cost of discovery of a deposit of given quality will necessarily rise over time. The improved technology may overpower any tendency for discovery costs to increase as geologists are forced to explore in less favorable areas.

Consider next the technology of exploitation. Technological progress can lead to wholesale shifts in the cost of exploiting different kinds of deposits, changing the quality ranking among them and invalidating the theoretical increases in royalties over time. At the turn of the century, the highest-quality deposits of copper consisted of small, high-grade veins;

the best iron ore deposits were those located close to the steel mills. Technological breakthroughs in mass-mining methods and in bulk transport made it much more economical to extract copper from more meager but much larger sulfide copper deposits and iron ore from landscapes of high-grade ores in faraway places such as Brazil and Australia. The royalty on the most highly valued deposits of earlier times fell to zero as they went out of use.

Clearly, the Hotelling rule has little relevance to the real world. Indeed, more recent advances in the theory of exhaustible resources, which take these complications into account, conclude that the royalty may rise or fall over time depending on the precise nature of the assumptions used (Fisher, 1981; Bohi and Toman, 1984). Nor is the Hotelling rule vindicated by empirical observation. Although royalties are difficult to measure, the tendency of mineral prices and mineral costs of production, in real terms, to fall over time suggests declining rather than rising royalties.[1] Hence, the presumption that the value of undeveloped resources in the ground tends to rise at the rate of interest has no general validity.

Nevertheless, the Hotelling rule has had and continues to have a profound impact on the theoretical and policy-oriented thinking about exhaustible resources in general and energy materials in particular (Aslaksen and Bjerkholt, 1986; Gordon et al., 1987; Hoel and Vislie, 1987; Morrison, 1987). Practitioners of resource policy have even asserted that the extracted-resource price would rise at the rate of interest. Such assertions, emanating undoubtedly from a perverted version of the Hotelling rule, explain the almost unanimous forecasts in the late 1970s and early 1980s about long-run increases in real petroleum prices by 2 to 3 percent per year (Exxon, 1980; U.S. Energy Information Administration, 1980; World Bank, 1981; Stobaugh, 1982).[2] Just like the original, the perverted Hotelling rule has no general support from theory or empirical observation.

---

[1] An analysis of prices is found in Grilli and Yang (1988). Analyses of long-run costs are found in Barnett and Morse (1963) and Barnett (1979).

[2] The consensus of oil price forecasts at the International Energy Workshops of the International Institute for Applied Systems Analysis (IIASA) in 1981, 1983, and 1985 further demonstrate the widespread use of the perverted Hotelling rule (see Adelman [1986]).

The empirical invalidity of Hotelling removes perhaps the strongest source of support for deferred mineral exploitation. With the real rate of interest at 2 percent and the value of the resource in the ground remaining stagnant, the present value of the income from exploitation in 20 years will be one-third less than that from exploitation today. The reasons for deferral must be strong to justify such a loss.

Saving for Future Generations

The argument that minerals should be left in the ground for future generations comes in two versions. The first and more nebulous, represented by the pleadings of the Club of Rome (Meadows et al., 1972), for example, sees a danger in the ever-rising global rates of exhaustible-resource exploitation and urges that more of the limited resource stock in the ground be saved to assure reasonable living standards for humankind in the distant future. This version has seldom had any significant impact on policy.

The second version is nationally focused. It is often encountered in countries that have made substantial progress in setting up a wide range of industrial activities but have yet to reach a mature stage of industrialization. It often starts out with the resource base that is known at the time and pleads that this should be preserved for the future needs of a growing domestic industry by restricting or prohibiting exports. Policies to restrict exports on this ground were implemented in Australia for iron ore in the 1940s (Adelman, 1970); in Canada for uranium in the 1970s (Radetzki, 1979) and for natural gas in the 1980s (Energy and Environmental Policy Center, 1985); in the Netherlands for natural gas in the 1970s (Estrada et al., 1988); and in Venezuela for iron ore in the 1970s (Radetzki, 1985).

Additions to Reserves

The argument that exploitation should be slowed or deferred to meet the needs of future generations suffers from two fallacies. In the first, the known resource stock is typically treated as a given, without appreciation for the full scope of its expansion through exploration and discovery. This oversight indicates a widespread misunderstanding of the mineral reserve concept, both among economic model builders and resource policymakers (Pindyck, 1978), manifested in a perennial concern about what will happen when the currently identified reserves have been used up.

In the early 1970s it was common to speculate that because the global reserves of many minerals constituted roughly a multiple of 30 of current annual production, severe problems of supply could be expected by the end of the century as the existing reserves were depleted. Today the reserve stock represents about the same multiple of current production, despite the sizable increase in output and the very large volumes extracted in the interim. Nevertheless, similar concerns about the future continue to be voiced, especially with regard to energy minerals.

The multiple of 30 is a number with considerable long-run stability throughout the history of industrialization. The explanation for this stability is that mineral reserves are much less given by nature than they are the result of conscious investment in exploration. Mining firms adjust their exploration efforts to keep reserves at a level that is comfortable for their long-run planning, but they have no incentive to invest in reserve creation above that level. A variety of events, such as unexpected luck in finding very large and rich mineral deposits or a burst of demand with rapidly expanding exploitation, can temporarily push the reserves-to-production multiple above or below the stable long-run level.

Very high elasticities in the reserve creation process can also be observed for individual nations, as the case of Australian iron ore illustrates. In 1939 the Australian government imposed an embargo on iron ore exports to prevent an early exhaustion of national reserves, then assessed at about 400 million metric tons (tonnes). Early in 1960, export restrictions were relaxed, following indications that the iron ore wealth in the country might be much larger than what had been identified. Export permission was granted for one-half of newly found iron ore (Adelman, 1970). After strong lobbying by the state government of Western Australia, the restrictions were abandoned late in 1960. There followed an explosive development of the many indications of iron ore resources in the state into massive identified reserves (Trocki, 1986).

So long as restrictions prevailed, there was no incentive to undertake detailed iron ore exploration. The domestic steel industry was located far away in the eastern part of the country. Its iron ore needs were small and could easily be satisfied from nearby iron ore mines that enjoyed a considerable competitive advantage because of their location. Hence, access to the world market was a precondition for the expansion of reserves. The futility and irrelevance of the earlier restrictions appear especially clear in retrospect: Between 1976 and 1986 alone, Australia exploited and exported more than 800 million tonnes of iron ore (*Mining Annual Review*, various issues), and yet by 1987 the total reserves were assessed at some 17 billion tonnes (U.S. Bureau of Mines, 1988). These exports were twice as large, and the reserves 42 times as large, as total Australian reserves in 1939.

Similar circumstances apply to uranium in Canada. In 1974 the Canadian government introduced policies requiring uranium-producing companies to set aside reserves adequate to cover 30 years of national requirements for existing and planned nuclear installations. It also refused approval for export contracts extending beyond 10 years, all with the stated intention of assuring domestic needs (Williams, 1976). Identified reserves in 1974 were assessed by the government at 360,000 tonnes, out of which 80,000 tonnes, or more than 20 percent, had to be set aside to satisfy the government policy. According to the government's analyses, existing reserves would not permit any Canadian exports beyond 1984. To assure future domestic needs in the way stipulated by the policy, Canadian miners would have to create and set aside reserves amounting to more than 800,000 tonnes before the end of the century. Even at the time it was introduced, this policy appeared exaggerated and unrealistic. Since then, the Canadian nuclear plans have been substantially scaled down, as have the plans of most other countries. At the same time, new and very rich discoveries have totally changed the Canadian uranium reserves picture. To the extent that it restricted exports, the policy damaged Canadian interests, because it limited the country's benefit from the extraordinary uranium boom in the latter half of the 1970s.

In comparison with mature industrialized nations where mineral exploration and exploitation have been going on for a long time, most developing countries constitute virgin territory for mineral explorers. This should create especially favorable prospects for additions to the existing reserve stock. What is more, an active and expanding mineral exploitation activity encourages exploration and increases the likelihood that reserves will grow. In the absence of large-scale mineral exploitation and exports over the past several decades, the reserves of iron ore in Brazil and Liberia, of copper in Chile, and of petroleum in Norway would certainly have been lower than the actual figures. There is no incentive to explore in the absence of reserve-consuming mineral production. Reserves tend to stagnate unless production is expanding. Hence, it is reasonable to claim that exploitation has expanded, rather than contracted, the level of reserves in these countries.

New Technology

The second fallacy of the argument that exploitation should be slowed or deferred for the benefit of future generations involves technical change. The preceding section on The Fallacies of the Hotelling Rule indicated how shifts in the technology of exploitation reduced the usefulness of mineral deposits that were highly valued in earlier periods. Sweden reaped very large benefits by exploiting its iron ore resources from the beginning of the century, when operations started, until the late 1950s. But the comparative advantage of these resources, based on their proximity to the steel mills in Central Europe, has gradually been lost because of the technological revolution in bulk transport. As a result, iron ore production in Sweden has lost its long-run viability. A decision early in the century to save the deposits for future use would therefore have reduced the income of earlier generations without yielding any benefit to the present generation.

Technological shifts of a similar nature are likely to occur in coming years as well. For example, a breakthrough in the technology of mining deep-sea nodules could render land-based deposits of nickel and cobalt worthless. Or, human inventiveness could result in the design of an inexpensive substitute for tin, sharply reducing the demand for that metal and making most of the world's tin deposits redundant.

Therefore, restricting mineral exploitation for the benefit of future generations is a potentially unproductive and risky undertaking. The interests of future generations are more likely to be served by a policy that sets aside the monetary proceeds from unrestricted exploitation today for their benefit, to grow at the rate of interest. Such a policy is likely to serve the best interests of the present generation as well.

Waiting for an Appropriate Institutional Framework

Sizable mineral activities were introduced into a number of developing countries before their independence or very soon after. At the time, their political systems either were in their infancy or were suppressed by colonial powers. The institutions and human capital to monitor or manage mineral ventures did not exist. Under these circumstances, mineral activity invariably was controlled by foreign multinationals, a large part of the mineral rent flowed abroad, and what remained in the country was often squandered by inexperienced administrators and public officials.

Such situations have provided a rationale for restricting mineral exploitation until the appropriate monitoring and controlling institutions have been set up and the government has acquired sufficient fiscal sophistication to extract part of the mineral rent and put it to wise use. Even more would be gained, according to this reasoning, if the deferral lasted until the human resources needed to operate the mineral industry without direct foreign involvement were in place, for then there would be no need to share the mineral rent with the foreign investors.

No doubt arguments of this kind have some validity in exceptional instances. The first Bougainville copper agreement in Papua New Guinea and the major agreements for mineral exploitation in Namibia come to mind. In both nations, the contracts were prepared by colonial administrations. They provided very advantageous fiscal terms to the foreign investors and offered little opportunity for the national development of management skills. The case of uranium mining in Niger, which dominated that country's national economy in the mid-1970s and suddenly endowed an unprepared government with large revenues, is also relevant. Human capital resources in these three countries were exceedingly scarce at the time, with very few university-trained nationals. It might be argued that deferring mineral development in Papua New Guinea and Namibia until independence, and slowing it down in Niger pending the development of a national administrative cadre, would have served the welfare of these countries.

Even in these exceptional cases, however, the case for delay is tenuous. In the cases of Papua New Guinea and Niger, for example, deferral would have meant that the extraordinary price booms of the 1973–1974 period for copper and 1976–1979 for uranium would have

47

been missed. The large size of the mineral rent reaped during these years may have more than compensated for the unfavorable contractual arrangements and government inexperience.

There are more general reasons why deferral may prove counterproductive. The monitoring and controlling institutions and the national managerial talent are unlikely to emerge in the absence of a clear need. Hence, a policy of deferring mineral extraction may also defer the establishment of the needed institutional and human capital assets—perhaps creating a vicious circle until the policy of deferral is discontinued. In contrast, a speedy start-up of exploitation may initiate a virtuous circle. The example of Papua New Guinea is again instructive. Despite its lack of experience, the national administration succeeded in renegotiating the exploitation contract soon after the start-up of production. The obsolescing-bargain hypothesis (see the paper by Moran in this volume) certainly provides a partial explanation of the renegotiation, but so does the experience gained from learning by doing. As a result, most of the rent generated by the project after 1974 accrued to the nation (Mikesell, 1975).

Similar arguments suggest that it may be futile to postpone mineral activity until a time when it can be managed without wholesale multinational involvement. Such involvement could in fact be a prerequisite for the initiatives to establish a national managerial cadre. The mineral rent that the foreigners take for the services they provide may appear excessive, but it can also be regarded as a temporary and unavoidable education expenditure.

Similarly, a lack of economic and financial sophistication in handling fast-growing revenue streams from the mineral activity is unlikely to be overcome other than by trial and error and may be quite costly. A national development plan to spend mineral revenues in a coherent way will not be worked out unless such revenues are projected. One must also distinguish between sophistication, which depends on extended interaction with modern management and technology, and wisdom, which does not. No sophistication is needed for the wise decision to deposit fast-growing mineral income in the bank, pending the emergence of sensible opportunities to spend the money. On the other hand, even a financially sophisticated government can squander the public income if it is unwise or dishonest or not concerned with social and economic development.

The conclusion that emerges from these arguments is that a developing country with no prior exposure to mineral activity has to incur a heavy setting-up cost to develop the institutions and human capital needed to reap significant national benefits from the exploitation of its mineral wealth. This setting-up cost seems unavoidable. Because it has to be incurred at some point, it does not constitute a justification for deferral.

One can, of course, argue that the setting-up cost will tend to decline with the level of economic development. Support for this view is provided by the speed with which countries at different levels of economic development have learned to manage their mineral industries efficiently after nationalization. The state takeovers of tin in Indonesia in the 1950s and of copper in Zambia in the 1960s were followed by decades of inefficient operations, largely because managerial cadres had to be built from the ground up. In Venezuela, in contrast, the ownership transfer of the iron ore industry in the 1970s was much smoother, mainly because the higher level of national economic development assured superior educational standards and national managerial exposure to various branches of industry (Radetzki, 1985). The relevance of this argument for deferring mineral exploitation is in serious doubt with regard to countries where mineral activities constitute the major potential for economic development. Deferral in such cases may prevent economic development from occurring at all.

This section has so far concentrated on the lack of national ability, defined in a broad sense, for dealing with the establishment and operation of a mineral sector. It also is necessary to mention a related, and apparently valid, argument for deferral. Investments in large mineral projects involve substantial efforts that can create strain, especially in a small, underdeveloped economy. At some volume of such investment activity, the cost to the national economy in terms of cultural disruption, inflation, an overheated labor market, and excessive reliance on imported factors of production will exceed the benefit of accelerated mineral income streams. Where the national absorptive capacity is limited in this sense, there is a strong case for deferral through sequential rather than simultaneous development when several profitable mineral investment opportunities emerge at the same time.

Deferral to Avoid Depressing Mineral Prices

Deferral of especially large projects has occasionally been recommended to avoid the detrimental impact the fall in market price caused by the added output would have on the host country. Two examples where such concerns have been raised are the huge Carajas iron ore project in Brazil, in production since the mid-1980s, and the Escondida copper mine in Chile, in operation in 1991.

Under the following assumptions about the Escondida venture, one might argue that deferral would have benefited Chile. World production of copper is taken at 6.5 million tonnes and Chile's production before Escondida at 1.2 million tonnes. Escondida will produce 0.3 million tonnes, equal to almost 5 percent of the world output and 25 percent of Chile's initial output.

Given its size, the operation of the new mine is expected to result in a price fall of 10 percent, from 80 cents to 72 cents per pound. The average total cost of the Escondida project, 50 cents per pound, will nevertheless make it a highly profitable operation. The profits will equal 22 cents per pound, or $146 million in total. However, the price fall will reduce the profits in Chile's other production—1.2 million tonnes—by 8 cents per pound, or a total of $212 million. So the returns to Chile's copper industry as a whole will decline by $66 million as a result of the proposed capacity expansion.

The results obtained from this reasonable set of circumstances lead, in the first instance, to the conclusion that the development of Escondida should be deferred. But will that conclusion stand closer scrutiny?

First, the assumption that a 5 percent addition to world supply will reduce prices by 10 percent is somewhat artificial. Those who argue the case appear to focus on how the additional output can be absorbed by the market without considering the impact of the new project on other suppliers. Hence, there is an implicit premise that the price elasticity of demand equals -0.5 but that the price elasticity of supply equals zero. The implicit price elasticity of demand seems reasonable if a period of two to three years is considered, given that the short-run price elasticities of demand have been measured at values between -0.1 and -0.2, whereas the long-run demand elasticities usually come out above absolute 1.0 (Fisher, Cootner, and Baily, 1972; Tan, 1987).

However, the assumption that the price elasticity of supply equals zero would make sense in the short run only in an extreme boom, when prior to the introduction of the new project the demand schedule intersects the supply schedule high on its vertical portion. In all other circumstances, one would expect a negative supply response from other sources, reducing the price impact. During recessions, when the demand schedule intersects the flat portion of the supply schedule, the price impact from additions to low-cost supply may be quite small. Yet, precisely because of the flatness of the supply curve, such additions will force numerous higher-cost producers out of business.

Second, for a mining project whose life is likely to extend over more than 20 years, it is essential to adopt an extended time perspective and consider what will happen beyond a 2- to 3-year time horizon. In a 5- to 10-year time perspective, the price elasticity of demand would probably be higher than absolute 0.5. More important, one has to consider not only what happens with the utilization of existing capacity, but also the extent to which other potential capacity will be brought into production if Escondida is not. In the case of copper and iron ore, the long-run supply schedule for potential capacity at the average prices that have prevailed in recent years is quite elastic, so the longer-term impact on price from deferring projects like Carajas or Escondida is likely to be very small.

Finally, the results obtained depend on the overall importance of Chile as a copper supplier and on the ensuing market power with which the country is endowed. Thus, the issue of deferring Escondida would not have arisen if Chile's original output was only 0.6 million tonnes, for then the profits from the new project would overwhelm the reduction in profits from existing facilities. The use of monopoly power to increase the benefits for Chile, if in fact such power is at hand, does not have to do with the size of the new project, but with supply manipulation in general. For example, the hypothesized figures indicate that Chile would benefit if it cut its output even below the pre-Escondida level. The full realm of monopoly power and how it relates to deferral are dealt with in the next section.

In conclusion, the arguments for deferring a mineral project because it is large appear to take an excessively restrictive and myopic view. Their validity weakens substantially when the time horizon is extended. Only in exceptional cases would a nation benefit from postponing large and profitable projects, and then the arguments seem to apply not so much to the size of projects as to the potential monopoly power of the country under scrutiny.

Delaying Projects to Strengthen Monopoly Profits

If a group of producing countries accounting for a sufficiently large share of total supply of a mineral commodity can agree to joint supply management, and if the absolute price elasticities of demand and of outside supply are low enough, the group may increase its mineral revenues and profits by cutting supply (Radetzki, 1975). Supply management through cartel action entails keeping the supply below what it would otherwise be. Consequently, if collusive action can benefit participating members by raising prices and revenues, it will pay to postpone capacity expansion even where it would have been profitable to develop in the absence of the cartel.

Theoretical reasoning indicates, and history confirms, that the viability of successful cartelization in this sense usually does not exceed three to five years. The cut in supply and the ensuing increase in price eventually encourage users to find substitutes for the commodity and independent producers to expand their supply. The price elasticities of demand and outside supply thus rise over time, reducing the cartel's market share and limiting its ability to reap benefits from the supply management scheme. Only in exceptional cases where the reaction lags are extended, the discount rates high, and the commodity belongs to the "indispensable," very-hard-to-substitute category, or where the cartel members possess a substantial competitive advantage based, for example, on a superior resource potential, will it be economical for cartel members to maintain prices above the competitive equilibrium.

The first case can be illustrated with a simple numerical example using constant money. For simplicity, the costs of production are disregarded. Assume that the producers use a discount rate of 10 percent and apply a time horizon of 10 years for their net present value calculations. Suppose that in the absence of supply management, their sales total $100 per year. The present value of the 10-year income stream amounts to $614. Now the producers restrict supply, and this raises their annual sales to $200 per year for the first 5 years, but the establishment of new production capacity outside their control reduces their sales to only $20 per year in the subsequent 5 years. The present value of the income stream under such supply management works out to $694, suggesting that the cartel is economically attractive to the participating members. Although the advantage is limited in time, the cartel nevertheless is worthwhile because it raises the net present value derived from the mineral activity.

The second case, involving a substantial competitive advantage based, for example, on superior resources, can be illustrated with the circumstances faced by the Middle East members of OPEC (Organization of Petroleum Exporting Countries).

In 1987, five OPEC members (Iraq, Iran, Kuwait, Qatar, and Saudi Arabia) had a total oil production capacity of 19 million barrels per day, but under the cartel's restrictive arrangements they together produced only 13 million barrels and sold it at prices of about $16 per barrel. Hence, their annual oil revenue totaled $76 billion.

If they wanted to, these countries could more than double their capacity and operate it profitably at prices below $1 per barrel (Adelman, 1986). The variable cost of production alone in many areas, including the continental United States, Canada, Alaska, and the North Sea, is several times higher. Restricted access to the vast and exceptionally rich resource base of the Middle East is the long-run modus operandi of the oil cartel. Deferral of new capacity in this area is deemed preferable, presumably because the cartel members believe that a capacity expansion of the order indicated would so depress prices that their total oil revenues would shrink. A detailed economic evaluation of a large-scale capacity and production expansion in the Middle East would depend on a number of assumptions about time horizons, demand growth, and the reactions to such a policy by other producers. It is quite possible that oil prices would eventually fall to $5 or less per barrel, resulting in a reduction of the Middle East producers' revenues from the 1987 level. In this light, abstention from capacity expansion may well be a rational policy choice.

Summary

Scrutiny of the various arguments suggesting that developing countries with favorable mineral resource potential would benefit by deferring mineral extraction finds such arguments have merit in only a few exceptional cases. The Hotelling rule, according to which the optimal drawdown of an exhaustible-resource stock in the ground raises its unit value at the rate of interest, typically provides strong support for deferral, but it holds only under a narrow set of assumptions rarely encountered in the real world. More realistic theories of exhaustible resources do not yield uniform conclusions about how the value of an undeveloped resource deposit will change over time.

The validity of the "saving for future generations" argument is refuted on several grounds. Mineral reserves can ordinarily be expanded with considerable ease, although misconceived policies to save a limited resource base for the future tend to discourage exploration and the discovery of new reserves. Highly valued mineral deposits can also become worthless as a result of technical change.

Developing countries without earlier exposure to minerals have to incur a one-time cost for setting up national human capital and institutions to monitor and manage the mineral activity. Until such infrastructure is in place, the rent generated by mineral extraction is likely to dissipate abroad or be wasted in other ways. However, the absence of such infrastructure does not provide a rationale for deferral. In fact, until a mineral sector has been established, there will be little incentive to build up the human and institutional resources needed to prevent the waste.

The argument against launching very large mineral projects that might be detrimental to the investing country because they depress prices is shown to be based on restrictive and myopic assumptions. The validity of this argument breaks down when dynamic considerations are taken into account and a longer time perspective is adopted. However, a shorter-run deferral, to avoid start-up during a mineral recession, may be warranted.

A valid case for deferred mineral expansion does exist if a country partakes in supply-reducing cartel arrangements that succeed in raising prices, revenues, and profits. However, both theory and history indicate that the necessary conditions for successful cartelization are quite rare.

There may also be a case for deferral through sequencing the development of several potential mineral investments, rather than developing them simultaneously, to avoid strain and disruption in the national economy of a developing country that would be caused by an excessive mineral investment program. Nonetheless, in most cases, a developing country would be best off by making use, without delay, of any profitable mineral resource potential it possesses.

Acknowledgments

Istvan Dobozi, Robert Drury, Roderick G. Eggert, Thomas D. Kaufmann, Robert H. Patrick, and John E. Tilton offered constructive comments, incorporated in this version of the paper.

References

Adelman, M. A. 1970. "Economics of Exploration for Petroleum and Other Minerals," *Geoexploration* vol. 8, no. 3/4, pp. 131–150.

———. 1986. "The Competitive Floor to World Oil Prices," *The Energy Journal* vol. 7, no. 4 (October) pp. 9–31.

Aslaksen, Julie, and Olav Bjerkholt. 1986. "Certainty Equivalence Methods in the Macroeconomic Management of Petroleum Resources," pp. 170–194 in J. Peter Neary and Sweder Van Wijnbergen, eds., *Natural Resources and the Macroeconomy* (Cambridge, Mass., MIT Press).

Barnett, Harold J. 1979. "Scarcity and Growth Revisited," in V. Kerry Smith, ed., *Scarcity and Growth Reconsidered* (Baltimore, Md., Johns Hopkins University Press for Resources for the Future).

Barnett, Harold J., and Chandler Morse. 1963. *Scarcity and Growth: The Economics of Resource Scarcity* (Baltimore, Md., Johns Hopkins University Press for Resources for the Future).

Bohi, D. R., and M. A. Toman. 1984. *Analyzing Nonrenewable Resource Supply* (Washington, D.C., Resources for the Future).

Energy and Environmental Policy Center. John F. Kennedy School of Government. 1985. "Prospects for Natural Gas Trade in North America and Western Europe." Discussion paper E85-08 (Cambridge, Mass., Harvard University, August).

Estrada, Javier, Helge Ole Bergeson, Arild Moe, and Anne Kristen Sydner. 1988. *Natural Gas in Europe: Markets, Organization and Politics* (London, Pinter Publishers).

Exxon. 1980. *World Energy Outlook* (New York, Exxon).

Fisher, A. C. 1981. *Resource and Environmental Economics* (London, Cambridge University Press).

Fisher, F. M., Paul H. Cootner, and Martin N. Baily. 1972. "An Econometric Model of the World Copper Industry," *Bell Journal of Economics and Management Science* (Autumn) pp. 568–609.

Gordon, R. B., T. C. Koopmans, W. D. Nordhaus, and B. J. Skinner. 1987. *Towards a New Iron Age? Quantitative Modeling of Resource Exhaustion* (Cambridge, Mass., Harvard University Press).

Grilli, E. R., and M. C. Yang. 1988. "Primary Commodity Prices, Manufactured Goods Prices and the Terms of Trade of Developing Countries: What the Long Run Shows," *World Bank Economic Review* vol. 2, no. 1, pp. 1–49.

Hoel, Michael, and Jon Vislie. 1987. "Bargaining, Bilateral Monopoly and Exhaustible Resources," in Rolf Golombek, Michael Hoel, and Jon Vislie, eds., *Natural Gas Markets and Contracts* (Amsterdam, North Holland).

Hotelling, Harold. 1931. "The Economics of Exhaustible Resources," *Journal of Political Economy* (April) pp. 137–175.

Kneese, A. V. 1989. "The Economics of Natural Resources," in Michael S. Teitelbaum and Jay M. Winters, eds., *Population and Resources in Western Intellectual Traditions* (New York, Population Council).

Meadows, D. H., Donelle Meadows, Dennis Meadows, Jorgen Randers, and William Behrens. 1972. *The Limits to Growth* (New York, Universe Books).

Mikesell, Raymond F. 1975. *Foreign Investment in Copper Mining: Case Studies of Mines in Peru and Papua New Guinea* (Baltimore, Md., Johns Hopkins University Press for Resources for the Future).

Morrison, Michael B. 1987. "The Price of Oil: Lower and Upper Bounds," *Energy Policy* (October) pp. 399–407.

Pindyck, Robert S. 1978. "Gains to Producers from the Cartelization of Exhaustible Resources," *The Review of Economics and Statistics* vol. 60, no. 2, pp. 238–248.

Radetzki, Marian. 1975. "The Potential for Monopolistic Commodity Pricing by Developing Countries," in G. K. Helleiner, ed., *A World Divided: The Less Developed Countries in the World Economy* (Cambridge, England, Cambridge University Press).

———. 1979. *Uranium: A Strategic Source of Energy* (London, Croom Helm).

———. 1985. *State Mineral Enterprises in Developing Countries: Their Impact on International Mineral Markets* (Washington, D.C., Resources for the Future).

Stobaugh, Robert B. 1982. "World Energy to the Year 2000," in Daniel Yergin and Martin Hildebrand, eds., *Global Insecurity: A Strategy for Energy and Economic Revival* (New York, Houghton Mifflin).

Tan, C. Suan. 1987. "An Econometric Analysis of the World Copper Market." World Bank Staff Commodity Working Paper 20 (Washington, D.C., World Bank).

Trocki, L. K. 1986. "An Analysis of the Role of Exploration in the Opening of New Iron Ore and Copper Mines" (M.S. thesis, Pennsylvania State University, University Park).

U.S. Bureau of Mines. 1988. *Mineral Commodities Summaries* (Washington, D.C., Government Printing Office).

U.S. Energy Information Administration. 1980. *1979 Annual Report to Congress* (Washington, D.C., Government Printing Office).

Williams, R. M. 1976. "Uranium Supply to 2000: Canada and the World." Paper presented at a meeting of the Geological Association of Canada, Edmonton, Alberta, May 15.

World Bank. 1981. *World Development Report: 1981* (Washington, D.C., September).

The Political Economy of Rent in Mining Countries

OLIVIER BOMSEL

Mining countries, for the purpose of this analysis, are those states where nonfuel mineral exports account for a large portion of total exports. This paper focuses on 15 developing countries, all with small populations and national incomes. Nonfuel mineral exports have accounted for 40 percent of the merchandise exports of these countries at least once since 1975. These 15 countries are Bolivia, Botswana, Chile, Guyana, Jamaica, Liberia, Mauritania, Morocco, Niger, Papua New Guinea, Peru, Suriname, Togo, Zaire, and Zambia. Guinea is also a mining country but is seldom considered in this analysis because of the lack of public data.

The exports of mining countries are far less diversified than those of large developing countries, such as India and Brazil, or those of mining-oriented industrialized countries, such as Australia and South Africa (figure 1). Therefore, mining and mineral processing are of vital importance to the mining countries. These activities also contribute a large share of the nation's import purchasing power and often are its largest industrial employer.

Despite their mineral wealth and the rents this wealth has generated, the mining countries have fared poorly in recent decades. Their ratio of foreign debt to gross domestic product (GDP) is among the highest in the world and in many cases debt continues to increase faster than GDP (figure 2). Per capita income, which with few exceptions

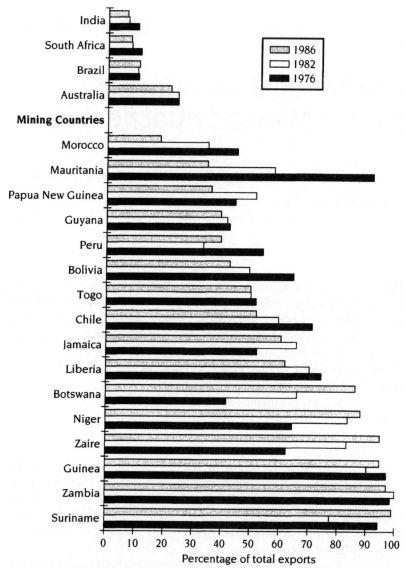

Figure 1. Nonfuel mineral commodity exports as a percentage of total exports for the mining countries and other selected states, 1976, 1982, and 1986.

Source: United Nations Conference on Trade and Development. *Handbook of International Trade and Development* (annual). Geneva.

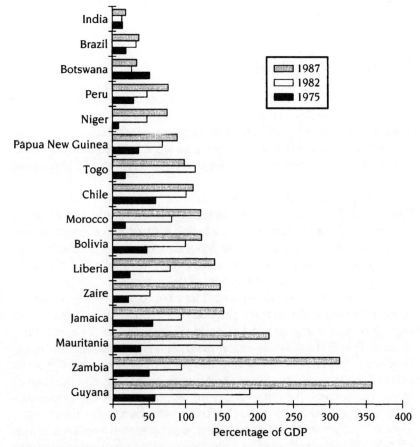

Figure 2. Foreign debt as a percentage of GDP for India, Brazil, and the mining countries, 1975, 1982, and 1987.
Sources: World Bank, *International Trade and Development Statistics* (annual); *World Debt Tables* (annual), Washington, D.C.

was very low to begin with, has actually fallen in more than three-quarters of the mining countries since 1975 (figure 3). For Zaire, Zambia, Guyana, Liberia, Peru, and Jamaica, the decline has been catastrophic.

This dismal record suggests that mining rents have done little to accelerate economic growth and may even have impeded the structural changes required for economic development. This paper explores this possibility by examining the use of mining rents and the consequences for the mining countries.

The Nature of Mining Rents

Economic rent is the surplus earned by factors of production over and above the minimum earnings necessary to induce their employment. Mining rent, in turn, is the surplus earned by a mineral deposit over and above the minimum earnings required to attract the capital, expertise, and other factors of production necessary to develop and exploit the deposit.

This surplus arises because mineral commodities are sold on international markets at prices reflecting global supply and demand, whereas their production costs, including a competitive rate of return on capital, depend on the mix of factors required for production and their prices. Consequently, production costs vary with the choice of extraction techniques and the quality of the deposit.

Mining rent can be separated into two components: (1) the difference between the market price and the production costs of the highest-cost or marginal producer and (2) cost differences among producers. The first, which is called absolute rent, is earned by every producer. This component of rent can be zero or even negative if the costs of the marginal producer are equal to or greater than the market price. The second, the differential rent, accrues only to the intramarginal producers, that is, to those producers with costs below those of the highest-cost producer. Differential rent is generated from deposits with particularly high-grade ores or from deposits that for other reasons are relatively inexpensive to exploit.

The mining rent a country earns can be measured by the difference between total production costs and the revenues received

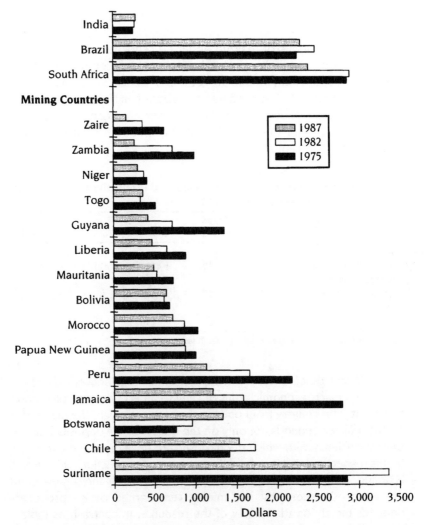

Figure 3. Per capita income for the mining countries and other selected states, 1975, 1982, and 1987 (in constant 1987 dollars).
Note: In a number of cases, the domestic currency was overvalued in the 1970s and undervalued in the 1980s relative to the dollar. This exaggerated the decline in per capita income expressed in dollars. Because of hyperinflation and a rapidly declining domestic currency, the exchange rate at the end of the year rather than the average for the year was used for Chile in 1982 and Peru in 1987.
Source: World Bank, *World Debt Tables* (annual), Washington, D.C.

(that is, the quantity produced times the market price). In 1988, for instance, when the price of copper averaged $1.30 per pound, the Chilean copper industry realized about 80 cents per pound in mining rent, or about $2.6 billion in total. The Zambian copper industry, in contrast, realized only about 30 cents per pound in mining rent, or $300 million in total. These estimates are based on the average variable production costs shown in table 1 plus estimated capital costs of 10 cents per pound for Chile and 5 cents per pound for Zambia, where production is less capital-intensive.

Table 1. Copper Output, Production Costs, and Productivity for the United States, Chile, Zaire, and Zambia, 1988

	United States	Chile	Zaire	Zambia
Output (thousands of tonnes)	1,438	1,451	471	450
Average variable costs (cents per pound)	53	40	40	95
Productivity (tonnes per worker-year)	60	43	13	7.5

Source: Estimates from the Centre d'Economie des Ressources Naturelles.

Mining rent is a transfer from consumers to producers, comparable from a macroeconomic point of view to an economic surplus transferred into the country from outside. In mining, though, the magnitude of this transfer depends not only on the quality of the deposits but also on the efficiency with which the deposits are developed and exploited. In this respect, mining is quite different from oil production. Once a well is drilled, labor productivity in oil extraction is roughly the same around the world. Consequently, the mining rent earned by oil producers depends mostly on the quality of the resource. In contrast, as table 1 shows, labor productivity in copper mining varies eightfold between Zambia and the United States, a difference that the lower cost of labor in Zambia can only partially explain. If the Zambian copper deposits were located in the United States, there is little doubt that they would be mined at lower costs and that the mining rent would be higher.

Economic development in the mining countries depends on both the generation and use of mining rents. Even when mining is in the hands of private firms and interests, the government has a critical role. It creates the fiscal regime and mining legislation that influence the magnitude and sharing of rents. It defines the rules that govern how mining rents are allocated and controls the distribution of rents to the public sector.

The rest of this paper examines how well the governments in the mining countries have performed these functions. The picture that emerges is distressing.

Earlier Studies

Two centuries after the Spanish conquest of the New World and the flow of gold into Europe that came in its wake, economists had already identified specific patterns associated with the absorption of mining rents. Renewed interest in this topic arose during the 1960s as conflicts between mining companies and host governments over the division of rents became widespread.

Although mining was widely recognized as a source of sectoral imbalances, it was generally accepted that mining rents could facilitate economic development by augmenting the stock of domestic capital (Hughes, 1975; Bosson and Varon, 1977). This positive view was tempered, however, by the Prebisch and Singer thesis that the gains in productivity obtained in mining and other primary product production are largely passed on to the consumer in the form of lower prices, whereas the prices of the manufactured goods imported by the mineral-exporting countries rise persistently (Singer, 1950, 1984; Prebisch, 1963). As a result, these countries suffer from a secular deterioration in the terms of trade that retards economic growth.

Looking back, it is far from clear that the terms of trade of the mining countries have deteriorated. Although the prices of mineral commodities have fluctuated, the prices of many manufactured goods, such as computers and other electronic goods, have fallen substantially.

In 1979 the World Bank conducted the first empirical study of the mineral-exporting countries (Nankani, 1979).[1] It found that the nonfuel

---

[1] This study compares the economic performance of two groups of mineral-exporting

mineral economies had enjoyed steady growth between the 1950s and the early 1970s, but that warning signs of pending trouble were already apparent. These countries suffered from (1) a technological and wage dualism, high unemployment rates, and low school enrollments; (2) a weakening of the agricultural and manufacturing sectors; (3) a growing trade deficit along with a shift toward imports favoring consumer goods at the expense of producer goods; (4) a steady increase in budget deficits; and (5) a higher rate of inflation than in the nonmineral economies. This study also highlighted the lack of linkages between the exporting sector and the rest of the economy and noted that the mineral rents accruing to the government bestowed on it a crucial economic role. The importance of reinvesting these rents was stressed, and governments were advised to promote saving and investment outside the mining sector. Wages in the mining sector and the exchange rate were identified as key variables in this process.

The need to control wages and the exchange rate and to manage the linkages between a booming mineral sector and the rest of the economy are similarly stressed by the Dutch Disease model, which also appeared in the late 1970s (Catz, 1977; Corden and Neary, 1982; Van Wijnbergen, 1984). Based on the experience of the Netherlands and its production of natural gas, this macroeconomic model contends that foreign exchange inflows from the booming mineral-export sector tend to raise the value of the domestic currency. In addition, the wages in the booming sector tend to increase, and these higher wages may spread to other economic sectors. The result is a loss of competitiveness for domestic producers outside the mineral sector.

According to the Dutch Disease model, however, the negative effects of a mineral boom can be reduced, or even avoided, if public policy keeps wages and the exchange rate from rising. The problem is that in many mining countries the political dynamics associated with the use of mineral rents make it difficult or impossible to institute such policies. This is perhaps best illustrated by the experiences of Zambia.

---

countries, one including and the other excluding exporters of mineral fuels, with a sample of countries that are not mineral exporters. Mineral-exporting countries are defined as countries for which nonfuel minerals represented more than 40 percent of merchandise exports and 10 percent of GDP between 1967 and 1975.

Mining Rents in Zambia

Rhodesian Selection Trust began large-scale copper mining in Zambia in the late 1920s. Since that time copper has accounted for a large share of total Zambian exports and government revenues. The mining rents flowing to the public sector, however, appear to have accentuated the structural imbalances within the country and undermined economic growth in four different ways.

1. Import Policy and the Agricultural Crisis Prior to Zambia's independence in 1964, only about 20 percent of the country's population lived in urban areas. Most agricultural activity was directed toward the production of food and oilseeds for domestic consumption, with maize accounting for 60 to 70 percent of the value of all marketed crops. Following independence, public policies were needed to strengthen the agricultural sector by providing peasants with secure property rights, suitable infrastructure and marketing facilities, an efficient credit system, and an accurate pricing mechanism. Unfortunately, the government failed to provide for these needs, and agricultural output, which had grown during the first years after independence, stagnated in the early 1970s. The year 1979 was particularly disastrous, owing to a drought and problems with the distribution of fertilizers and agricultural credit (World Bank, 1981).

To make up the deficit in domestic food production, the government encouraged food imports. During the 1970s, food from abroad cost the country about $50 million annually, which was 6 to 10 percent of total visible imports and about 40 percent of the value of all marketed crops. In 1979, when maize production dropped by 50 percent, maize imports alone jumped to $75 million.

Although food imports did not consume an unduly large share of Zambia's import purchasing power, they had a disastrous effect on domestic agriculture. Low food prices combined with subsidized food imports made it impossible for many peasants to remain in farming. The dynamics of the government's import policy and the agricultural crisis it precipitated are shown by the arrows representing "import policy effect" in figure 4.

2. The Redistribution of Income The Zambian crisis in agriculture increased the pressure on peasants to migrate to urban areas,

68

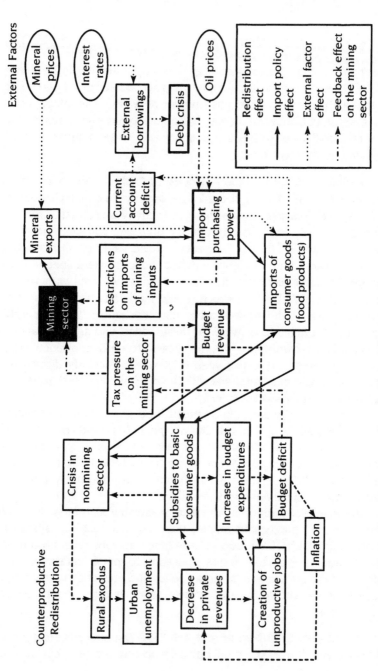

Figure 4. Economic dynamics of rent: a rent-earning sector financing structural imbalances.

particularly to Lusaka and the Copperbelt provinces. Some 94,000 people per year abandoned the countryside for the cities between 1969 and 1974. That number rose to 125,000 per year between 1974 and 1979. About half the population now lives in urban districts. The sharp income differentials between industry workers and civil servants on the one hand and peasants on the other, combined with the availability of price-controlled goods in urban areas, have stimulated the migrations.

To cope with growing urban unemployment, the government has created unnecessary jobs in the public sector and underwritten the price of many consumer goods including food. The agricultural subsidies, which by the late 1970s were equivalent to 7 percent of GDP, lowered food prices for the urban consumer but failed to raise farm income and so stimulated the rural exodus. The counterproductive effects of these government policies to redistribute income toward the urban centers are shown by the arrows representing "redistribution effect" in figure 4.

3. Vulnerability to World Market Forces Revenues from the mining sector allowed the Zambian government to set up these forms of income redistribution, none of which create goods or in any other way add to the economic surplus of the domestic production system. While aggravating the difficulties of agriculture, the allocation of mining rents led to ever-larger government expenditures that made balancing the budget more and more dependent on copper revenues.

Zambia, as a result, became more vulnerable to the vicissitudes of international markets. The sharp drop in copper prices in 1975 following the first oil shock caused a 50 percent slump in the country's terms of trade. Domestic consumption was maintained, primarily through deficit financing, on the expectation that the drop would be temporary. The government budget and external account, both in surplus during the boom years, moved into large deficits (figure 5).

At the same time, the government sought new sources of funds. It placed increasing pressure on the mining sector for a greater share of the mining rents. Foreign loans reinforced the additional funds taken from the mining sector. With the external current account deficit averaging 16 percent of GDP between 1975 and 1987, Zambia's foreign debt rose exponentially. The increase in interest rates in the early 1980s

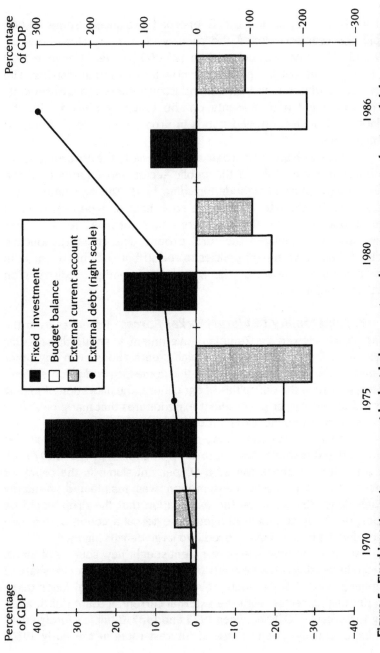

Figure 5. Fixed investment, government budget balance, external current account, and external debt as a percentage of GDP for Zambia, 1970, 1975, 1980, and 1986.
Source: International Monetary Fund.

increased the country's debt-service payments. Zambia has experienced six debt-rescheduling agreements and is still negotiating for further adjustments. Its debt service is now heavily concentrated on interest payments, while the principal continues to grow faster than GDP. As a result, the country's foreign debt is now more than three times GDP (figure 2). The arrows representing "external factor effect" in figure 4 highlight the ways in which the Zambian economy has grown more vulnerable to external market forces.

4. Effects on Mining   For a number of reasons, including a desire to increase its share of the mining rent, the Zambian government, after independence, extended its control over the copper companies operating in Zambia. In 1969 the industry was partially nationalized, with the government agreeing to purchase over a 10-year period 51 percent of the outstanding shares of the copper companies at a cost of $331 million. The industry was then consolidated into two groups. The operations of Anglo-American formed Nchanga Consolidated Copper Mines Limited (NCCM), and those of Roan Selection Trust International, a wholly owned subsidiary of Amax, became Roan Consolidated Mines Limited (RCM). The government formed a holding company, the Zambia Industrial Mining Corporation Limited (ZIMCO), which held the government's share of both companies. Anglo-American and Amax continued managing their respective companies until 1974, when the government terminated their management and marketing agreements in return for a lump sum payment.

The boom in copper prices in 1974 probably encouraged the government to increase its control over the two companies. However, when copper prices fell in 1975, the financial situation of both deteriorated. As the need for more equity became apparent, the government approached Amax and Anglo-American about investing new funds. Neither was interested, and in 1979 the government increased its shareholding in both companies. It then announced the merger of RCM and NCCM in May 1982 to form Zambian Consolidated Copper Mines Limited (ZCCM). Thus, the final step in the nationalization of the Zambian copper industry occurred as a result of the withdrawal of foreign operators in response to growing pressures from the Zambian government.

The withdrawal of the foreign firms, combined with the decline in foreign exchange earnings and the lack of domestic resources, led to insufficient investment in the mining industry. At the same time, the rural exodus toward the Copperbelt stimulated unproductive employment within the copper sector. The number of people employed in mining jumped from 52,000 to 64,000 between 1965 and 1975 and has remained at this high level even though copper production has declined by 40 percent. About 3 million people now depend on ZCCM salaries in the Copperbelt.

Just how the allocation of mining rents over time has threatened the country's copper industry and consequently the very source of those rents is illustrated by the arrows representing "feedback effect on the mining sector" in figure 4.

Mining Rents in Other Countries

An examination of other mining countries shows that the Zambian experience reflects several common patterns: first, the use of mining rents to avoid or postpone needed, but unpopular, structural adjustments in agriculture or other important economic sectors; second, the use of mining rents for economically unproductive welfare purposes; and third, the use of rents in ways that increase the dependence and vulnerability of the domestic economy to external forces.

Avoiding Structural Adjustments

A few examples will illustrate that Zambia is simply part of a well-established tradition of mining countries using their rents to postpone needed structural adjustments.

Chile in the 1870s Despite the substantial government revenues and export earnings generated during the 1870s by the Chilean copper industry, the largest in the world at the time, little was done to help domestic producers adjust to the growing threat emerging from the development of North American mines using new mechanized techniques. As a consequence, copper mining in Chile had all but disappeared by the end of the 1880s.

The government responded by issuing permits to British companies to develop nitrate deposits found on its borders with Bolivia and

Peru. The three countries collided in their rush to capture the mining rents, producing the Pacific War of 1879 to 1882, also known as the Nitrate War. Chile's victory allowed the country to substitute the mining rents from nitrates for those lost from copper. Nitrate exports accounted for nearly 25 percent of Chile's GDP between 1880 and 1904, permitting the government to sustain domestic consumption by importing consumer goods. As a result, the domestic production of consumer goods stagnated. This policy collapsed with the disruption of the world nitrate market in the early 1920s.

Chile in the Late 1940s   The import substitution policy initiated in the mid-1930s by the government had stalled. Production and productivity had stagnated and even declined, while the gap between domestic and world prices widened substantially. The government responded by raising the taxes on the copper mining companies controlled by U.S. corporations. Until 1955, when these corporations threatened to withdraw from Chile, rising mining rents financed unproductive subsidies for domestic manufacturing.

Guinea in the 1960s and 1970s   When agrarian reform failed to revive Guinea's ailing agricultural sector in the early 1960s, the state responded by developing the country's mining sector. The exploitation of the bauxite deposits of Kindia with the Russians, and of Boké with a consortium of Western aluminum companies, allowed the government to pay for food imports that offset the decline in domestic agricultural output.

The continuing decline in agriculture, coupled with rising food imports, eventually prompted the Guinean government to increase its share of the mining rent. In 1974, following the lead of Jamaica, it imposed a new levy on bauxite exports.

Many other examples could be cited of mining countries using their mining rents to postpone the difficult political decisions and the painful economic adjustments needed to resolve pressing problems in agriculture, manufacturing, and other important sectors. The use of rents to finance imports to compensate for declining output in an ailing sector, although common, has had particularly onerous consequences, because this policy tends to aggravate rather than alleviate the problems of the troubled sector. Mining rents have in this way prolonged the procrastination and agony caused by economic problems, allowing governments to demure from making unpopular but needed changes.

Unproductive Welfare Programs

The decline of agriculture or other traditional productive sectors creates economic hardship for that part of the population that has historically made its living from the ailing sector. If peasants cannot feed themselves, if industrial workers cannot survive on the going wage, or if unemployment affects too large a share of the working force, the government is under tremendous pressure to provide some kind of income assistance to those in need. The experiences of mining countries show that this assistance is provided in a variety of ways: unproductive job creation in the public sector, subsidies to the private manufacturing sector, permission for state enterprises to operate at persistent losses, and the use of public funds to maintain artificially low prices for important consumer goods.

Revenues from the mining sector, which reflect capital inflows from the rest of the world, have thus financed the creation of uneconomical income redistribution systems—uneconomical in the sense that they do not encourage the production of the optimal mix of goods and services or the use of the most efficient production techniques. Indeed, this policy has often increased the problems of domestic producers, as the Zambian subsidies on food imports illustrate. In the rest of the world, where mining rents are negligible or relatively much less important, wages, productivity, and domestic prices are more closely connected.

Vulnerability to External Forces

In many mining countries the use of mining rents for welfare programs, which, once in place, are politically very difficult to cut back or eliminate, has increased the dependence of the domestic economy on these capital transfers from abroad. Moreover, pressure to expand these programs has forced governments to seek more and more funding.

Their efforts are normally first directed at obtaining a larger share of the total mining rent. During the 1960s and 1970s, the conflict incited by these efforts deterred new investments from abroad and led to nationalizations and the withdrawal of foreign companies in a number of mining countries. In addition, governments borrowed heavily from abroad during the 1970s to reinforce the transfers from the mining sector. The mining countries are now among the most indebted in the world.

Furthermore, in some countries—Peru, Zaire, Zambia, Liberia, and Bolivia, for example—the relentless quest for new and greater transfers has ultimately threatened the viability of domestic mining. In spite of diminished profits, state-owned enterprises have continued to pay the state an increasing share of their hard currency earnings. In some instances, mining companies have even acted as shadow borrowers to enable the government to obtain new international loans. As a result, a number of these firms, including Gecamines in Zaire, Centromin in Peru, Comibol in Bolivia, and ZCCM in Zambia, have encountered production problems owing to underinvestment and the lack of spare parts and other imported goods.

There are exceptions. The mining sectors in Morocco and in Chile after 1975 were not only maintained but expanded. At the same time, these countries were trying to diversify their exports. Mining rents, although rising, were not sufficient to finance current budget expenditures, requiring an increase in foreign borrowing. Net capital inflows into Chile through the banking sector, for example, amounted to $2 billion per year between 1976 and 1982.

Thus, the use of mining rents to compensate for structural problems and to provide welfare assistance has left the mining countries more vulnerable to the vicissitudes of the international markets for capital and mineral commodities. It is ironic that so many of the mining countries have lost much of their economic independence only a decade or two after winning political independence.

Prospects

This analysis suggests that much of the blame for the poor economic performance of the mining countries must be placed on the way these countries have used their mining rents. A destructive pattern emerges that encompasses the four "vicious-flow" networks illustrated in figure 4. Domestic economic problems cause rising prices for food and other consumer goods. Responding to public pressure, governments subsidize low-cost imports. This seemingly benevolent policy sets in motion a series of developments that leave the country with high urban unemployment, expensive and unproductive welfare programs, balance-of-payments problems, large foreign debts, a weakened mining sector, and a domestic economy much more dependent on the rest of the world.

When international or domestic conditions retard or reverse the capital flows pouring into the mining countries through the mining and banking sectors, a severe economic recession usually results. In Chile, for example, the reversal of financial flows in 1982 produced a slump in internal demand and the collapse of a large part of the manufacturing sector. With the help of the World Bank and other international organizations, many mining countries are now trying to restructure their economies to restore economic balance.

The perverse dynamics associated with mining rent produces stagnant or declining per capita income. Mineral wealth, which should be contributing to economic development, is actually inhibiting the public policies and private activities necessary for growth.

To extricate themselves from their current morass, all mining countries have in the short run had to reschedule the servicing of their foreign debts. Over the longer run, however, they must also rebuild their productive sectors harmed by the past allocation of rent.

Before mining rents can be used constructively to stimulate economic growth, new links must be forged between income distribution and productivity. Although foreign debt is a major constraint on growth over the near term, because it obliges many mining countries to transfer to creditors much of their export surplus, the heart of the problem lies elsewhere. The critical issue is how these countries can productively use their underemployed or unemployed labor resources in conjunction with their mining rents to accelerate domestic growth.

This question has no simple answer. It is clearly not appropriate to recommend general policy measures based on macroeconomic principles assuming that governments, after a string of serious errors, have suddenly become wiser and less susceptible to the political pressures perverting public policies in the past. The issue, therefore, is not to define general policy measures but to better understand where needed reforms might begin.

Chile and Morocco perhaps provide examples. In recent years, these two countries have boosted their mineral exports and are now trying to develop the export potential of their agricultural resources. They have also managed to keep their foreign debt from rising faster than GDP and are witnessing the emergence of a small but important group of national entrepreneurs.

Despite these successes, however, both countries continue to suffer from high unemployment and remain heavily dependent on imports of consumer goods. Because of continuing weaknesses in traditional agriculture and small-scale manufacturing, they are still facing major income distribution problems that weaken the linkages between wages and productivity.

For sub-Saharan Africa, diversifying exports is much more difficult because few of the countries in this region possess significant export opportunities outside of mining. The substitution of domestic food products for imports will also be difficult because, in most countries, the commercial agricultural sector has suffered severely from inept public policies. It will not be easy to persuade the peasants who have left the countryside for the major cities to return. Even if these countries manage to restore their trade balances and to keep their ratio of debt to GDP from rising (thanks to higher mineral prices), they will encounter serious barriers in their efforts to establish new production sectors to revive economic growth.

Moreover, in many of these countries the counterproductive wage policy has even reached the mining sector itself, weakening the future contribution it can make to growth. Some major mining companies, such as Centromin in Peru and ZCCM in Zambia, are as much concerned about providing jobs (ZCCM employs 60,000 people directly and provides income, directly and indirectly, for some 3 million people) as producing and selling copper or other mineral commodities. Between 1982 and 1986, while U.S. producers increased their labor productivity by 50 percent, Zambia adapted to depressed copper prices by devaluing its currency. It is nearly impossible for some mining countries to maintain high productivity in the export sector when productivity elsewhere in the economy is so low. The risk, as a result, is that these countries could lose their competitiveness in mining and mineral production.

In some countries where economic policies have badly damaged such traditional economic sectors as agriculture (again, Zambia provides a good example), the government is composed of the people who have devised and implemented these policies. As the largest employer within the country, the state and its employees constitute the strongest pressure group for maintaining the current unproductive income distribution system. Moreover, few government employees possess the technical and managerial abilities necessary to develop meaningful reforms and to rebuild the economy.

Paradoxically, the state-owned mining company is often the only institution that possesses people with these talents. It may be the only place where new policies aimed at raising productivity can arise. Such policies, however, inevitably collide with the perceived interests of workers and their unions and hence need strong government support. But government officials, having benefited for years from the income distribution system on which they built their political power, may not possess the political strength to support the needed reforms.

References

Bosson, Rex, and Bension Varon. 1977. *The Mining Industry and the Developing Countries* (New York, Oxford University Press for the World Bank).

Catz, P. 1977. "The Dutch Disease," *Management Today* (March) pp. 78–81.

Corden, Max, and J. Peter Neary. 1982. "Booming Sector and Deindustrialisation in a Small Open Economy," *Economic Journal* vol. 92, no. 368 (December) pp. 825–848.

Hughes, Helen. 1975. "Economic Rents, the Distribution of Gains from Mineral Exploitation, and Mineral Development Policy," *World Development* vol. 3, no. 11/12, pp. 811–825.

Nankani, Gobind. 1979. "Development Problems of Mineral Exporting Countries." World Bank Staff Paper 354, August (Washington, D.C., World Bank).

Prebisch, Raul. 1963. *Towards a Dynamic Development Policy for Latin America* (New York, United Nations).

Singer, Hans W. 1950. "The Distribution of Gains Between Investing and Borrowing Countries," *American Economic Review, Papers and Proceedings* vol. 40, no. 2 (May) pp. 473–485.

———. 1984. "Terms of Trade Controversy and the Evolution of Soft Financing: Early Years in the UN, 1947–51," in Gerald M. Meier and Dudley Seers, eds., *Pioneers in Development* (Oxford University Press for the World Bank).

Van Wijnbergen, Sweder. 1984. "The Dutch Disease: A Disease After All?" *Economic Journal* vol. 94, no. 373 (May) pp. 233–250.

World Bank. 1981. *Zambia: Country Economic Memorandum*. World Bank Report 3007-ZA (Washington, D.C., World Bank).

Economic Policy in Mineral-Exporting Countries: What Have We Learned?

PHILIP DANIEL

In the half-century since World War II, mineral-exporting countries have experienced a series of massive swings in their terms of trade that have amounted to severe macroeconomic shocks. Whether booms or slumps, the shocks have had a profound impact on the structure and welfare of the economies and societies concerned.

From the vantage point of North America or Europe, the booms attracted the most attention: the peaks in nonfuel mineral prices during the Korean and Vietnam wars and the oil shocks of the 1973–1974 and 1979–1980 periods; both were associated with temporary peaks in non-fuel mineral prices as well. The slumps that followed were perceived as welcome respites. But for lower-income commodity exporters, they presented a major challenge to the social fabric, the conduct of economic policy, and the basic task of meeting the material needs of ever-expanding populations. The slumps in nonfuel mineral prices following the Korean and Vietnam war booms were perhaps foreseeable, but the sudden collapse of mineral prices after 1974, the prolonged depression in mineral markets during the first half of the 1980s, and the oil price collapse in 1986 were much more difficult to predict.

This paper addresses how mineral revenues, with their associated cycles and shocks, are moved through the economies of lower-income mineral exporters and asks whether enough has been learned about the processes involved to make appropriate economic policy responses clearer for the future. In so doing, it explores why mineral exports from lower-income countries so often seem to be associated with a retardation of growth and development instead of with a relaxation of foreign exchange, savings, and government revenue constraints and thus a contribution to growth and development.

A five-part approach is taken: (1) some important characteristics of mineral production are outlined, and three interdependent categories of policy choice are introduced; (2) theoretical approaches from the tradition of development economics are reviewed to help illuminate the issues; (3) the record of economic performance in mineral-exporting countries is considered; (4) the economic policy choices to be made in absorbing mineral revenues are examined; and (5) some unsophisticated rules that may be of use in increasing the developmental benefits of mineral exports are offered.

Mineral Market Characteristics and Economic Policy

Booms or slumps in mineral prices produce variations in the availability of resource rent. Resource rent can be defined as the value of the product of a mineral resource minus all the costs of production, including the minimum returns to capital that are necessary to induce investment, including exploration investment (Garnaut and Clunies Ross, 1983). Rent is thus a stream of income not required to reward capital or labor (reproducible factors of production) or to pay for intermediate inputs. Hence, production linkages (inputs to mineral industries from other domestic sectors or, alternatively, from mineral industries to other domestic sectors) are, by definition, low in rent-producing activities because the portion of final product value committed to intermediate inputs and reproducible factors is relatively low (Gelb and Associates, 1988, p. 15). Rent in this sense is properly referred to as "pure" or Ricardian rent and distinguished from the concept of "scarcity" rent associated with the exhaustible nature of mineral resources.[1] Ricardian rent arises from resources of

---

[1] The notion of scarcity rent, described more fully in the paper in this volume by Radetzki, is derived from the Hotelling rule: In equilibrium (and under other restrictive conditions

superior quality (which need not be exhaustible, but only in fixed supply—hence the derivation of the concept from the pure rent of land) where the price of the product is independent of the cost of exploiting the resource.

Market Characteristics of Mineral Commodities

With market prices set by the interaction of aggregate international supply and demand, there is no necessary connection between the costs of production of an individual mine or oilfield and the market price obtainable for the product. Potential resource rent from an individual deposit will vary according to the deposit's cost-competitiveness with other deposits of the mineral. This position will be affected by many factors—including the richness of the ore, accessibility of the mine site, and cost of infrastructure—that will determine the availability of resource rent in the face of a given average product price over the life of an individual deposit.

Mineral prices, however, are liable to wide fluctuations. Sharp price increases will accentuate booms in countries with rich mineral deposits and, conversely, rapid price falls will reduce or eliminate rent for a time. The basic causes of instability in mineral markets are well known. When demand increases and existing mines approach full capacity, even large increases in price cannot induce a significant short-term increase in supply because new capacity can take years to develop. Demand is very responsive to changes in the incomes of consumers (represented by the level of economic activity in industrial user

---

governing the operation of asset markets), the unit price of a nonrenewable resource will rise at the real interest rate because the resource will otherwise be extracted and the proceeds used to acquire interest-bearing assets.

In the real world, the closest counterpart to scarcity rent is the additional return investors in minerals seek to compensate for the expected cost of finding new deposits of an increasingly scarce resource. In the Ricardian sense, this is not rent at all, but part of the market-determined rate of return required to secure the initial commitment of capital.

This distinction helps to resolve a common dispute between investors in the minerals industry and host governments. Ricardian rent is properly taxable by the resource-owning authority, whereas the "rent" which represents the market-determined expectation of the cost of finding new deposits (the sum of successful and unsuccessful exploration and development) will be part of the required return to capital.

countries), but in the short run it is not responsive to mineral price changes because substitution of cheaper materials takes technical change, new investment, and time.

Where prices are set in competitive markets without supply control arrangements among producers (as in copper, gold, and, increasingly, aluminum), market changes produce volatile prices. Where prices are mainly set by producers, often under state control and with formal or informal collaboration arrangements among those producers (iron and steel, diamonds, and at times, petroleum and tin), instability will appear in the volume of sales. In either case, there will be pronounced instability in the revenues of producers.

Mineral market instability appears to have increased over time (United Nations, 1982; Tilton, 1981). For one thing, in the United States at least, the behavior of stocks became less stabilizing in the 1960s and 1970s. Also, combined industrial production of the metal-consuming countries became subject to wider fluctuations, at least until the sustained industrial recovery of the 1980s, and the business cycles of the major industrial countries became more closely synchronized. Further, there was an increase during the 1960s and 1970s in state ownership and control of mining sectors that may have contributed to instability (Radetzki, 1985). If, for social reasons (including overvalued exchange rates and excessive debt burdens), output is not curtailed when prices fall, then the downswing will be accentuated and instability enhanced. Finally, currency instability and the proliferation of instruments for investment in commodity trading may have attracted increased flows of speculative funds into commodity markets.

The 1980s saw heightened potential for instability in at least three minerals markets that had enjoyed a substantial degree of supply control by producer interests: (1) diamonds, where prices and sales fell sharply in the 1981–1982 period; (2) tin, with the failure of the International Tin Council in the 1985–1986 period; and (3) petroleum, where the strength of OPEC (the Organization of Petroleum Exporting Countries) was severely tested after 1985. All these markets came under pressure in the wake of a prolonged slump and oversupply in other mineral markets while the industrial countries of the West adopted tight monetary policies to arrest inflation.

The instability of resource rent has also been accentuated by changes in the price of inputs as well as outputs. Mineral exporters have faced sudden price movements, such as the oil or interest rate "shocks," in the same way as the rest of the world.

Mineral Revenues and Economic Policy

The magnitude of resource rent from a particular mineral deposit cannot be determined in advance. It depends on the evolution of costs at the mine and on the course of prices for the product in international markets and, thus, on all those factors that determine these costs and prices, including the cost of competing deposits—both those in production and those yet to be discovered.

The government of a mineral-exporting country faces a three-part problem: (1) how to identify, maximize, and retain mineral rent for domestic saving or consumption (the taxation problem); (2) how to manage economic activity over time in the face of risk and uncertainty about the rent receipts (the macroeconomic problem); and (3) how to distribute mineral revenues across uses and over time (the absorption problem). It faces risk in the sense that its predictions may turn out to be wrong, and uncertainty in the sense that it faces certain parameters or events whose values are inherently unknowable. It faces cyclical and structural changes in the outside world and may on occasion have considerable difficulty distinguishing between the two.

If the amplitude of cycles in mineral markets has increased (which is one meaning of an increase in instability), and if the cycles themselves have become more prone to interruption by shocks, then the degree of risk and uncertainty facing mineral exporters will have increased, as will the variability of realized rent receipts. With unstable prices but relatively stable trends in costs (including returns to capital), mineral rent is in any case the most unstable component of final product value.

Simply put, governments in mineral-exporting countries find themselves in the position of seeking to maximize windfall receipts from mineral exports and then confronting the economic consequences of the alternate presence and absence of these same windfalls.

The taxation problem consists of the design, in the face of uncertainty, of a system of revenue sharing between mining companies and the government that maximizes the flow of government revenue over time. If the tax system deters exploration and development activity that would otherwise be economically justified, potential rents will remain unrealized. If it causes a resource to be exploited inefficiently, rent will be dissipated. If it leaves substantial portions of rent to accrue to recipients other than the state (whether because it is poorly designed or because there is permanent or temporary monopoly in the supply of some factor of production), then rent is diverted (Garnaut and Clunies Ross, 1983, pp. 188–196).

The macroeconomic problem involves maintaining the traditional targets of external and internal balance. External balance prevails where the net accumulation of foreign liabilities by the government and the banking system is not accelerating faster than the expected servicing capacity of the economy and where adequate foreign exchange reserves exist for external payments requirements throughout any external economic cycle the economy faces. External balance is desirable for two reasons: (1) because its absence implies a reduction in the consumption and welfare of future generations, and (2) because other things being equal, an absence of controls on current transactions and an absence of parallel (unofficial or illegal) markets are preferable to their presence.

Internal balance is a difficult notion in a low-income economy. To form a working definition, employment and price stability must be separated. Full employment of resources makes a greater contribution to welfare than less-than-full employment does. Internal imbalance often shows in the pace of migration from country to town; if migration takes place faster than the demand for urban labor expands, then there is prima facie evidence of internal imbalance. As for price stability, a gently rising price level is preferable to hyperinflation. Finally, generally stable economic activity reduces the risk premiums that investors will attach when making decisions, helps avoid disruption of public- and private-sector projects, and provides a secure position from which to negotiate or trade with the world.

The absence of external and internal balance raises problems of equity as well as of efficiency. Instability, rationing, controls, rent seeking, and so on create an environment where those without assets and know-how tend not to benefit. Moreover, financial and monetary instability jeopardizes any major social transformation or development program.

Sound macroeconomic management is a necessary but not sufficient condition for the achievement of a development path of growth and transformation.

The absorption problem—or how to "sow" the oil (gold, copper, . . .)—is perhaps the toughest problem of all. Again, the policy objectives are traditional: to maximize the level of consumption and to optimize its distribution over time. The instruments and choices for doing so, however, are many and complex, and the outcomes are not independent of the solutions chosen for the taxation and macroeconomic problems.

Theoretical Approaches

The conundrum posed by the experience of mineral exporters is how a free flow of foreign exchange (usually in the form of additional public revenue) can fail to relax development constraints and thus contribute to the long-term improvement of consumption and welfare. Fluctuations in revenue are not necessarily damaging to growth prospects, and the need for wise deployment of the proceeds of a nonrenewable resource can be foreseen.

Growth and Gaps

The expansion of exports and the inflow of foreign investment to create export industries have traditionally been regarded as beneficial to a developing economy. Domestic market size is not a constraint on export growth, and the diversion of domestic resources to export production, where this exploits the country's comparative advantage (usually the case for mineral exports), permits the importation of other goods from lower-cost international sources and, thus, raises real income. The standard theory of comparative advantage points to gains from trade in the form of higher income generated by greater efficiency in the allocation of given domestic resources. Foreign investment in export industries, moreover, may supplement domestic savings and encourage a process of technology transfer. During the colonial era, primary export industries were presumed to increase the domestic propensity to save and to stimulate new wants and aspirations by encouraging widespread monetization and the importation of previously unknown goods and services (a phenomenon that could also reduce domestic savings).

In neoclassical growth theory, growth occurs by expanding the frontier of combinations of available factors of production. Improvements come by increasing the quantity of factors and the efficiency of their allocation. In the simple two-factor model, growth depends on increases in the labor force and in capital formation. If labor is abundant—the typical situation in low-income countries—then domestic savings are the constraint on capital formation and, hence, growth. Where labor and other domestic inputs are in ready supply but export production is limited by scarcity of importable inputs, the availability of foreign exchange becomes the second constraint—the two-gap model underlying the expansion of foreign aid in the 1950s and 1960s. Where the public sector plays a major direct role in capital formation, or where public infrastructure is needed to stimulate private investment, government revenues become a third constraint or gap.

Mineral revenues, like foreign aid, offer the opportunity to close all three gaps. Therefore, from the standpoint of neoclassical growth theory on international trade and growth, windfall revenues from mineral exports should be a blessing.

Structuralism and Linkages

As decolonization proceeded, and as economic nationalism took its place and development became both an objective and a subject of study in its own right, the neoclassical orthodoxy came under fire. The principal criticisms were (1) that primary export industries tend to function as enclaves with very limited linkages to other sectors of the local economy; (2) that the resulting imports tend to stifle local industry and prevent the realization of "dynamic efficiency" gains from industrialization, while export industries draw scarce capital and entrepreneurship away from production for the domestic market; and (3) that with foreign investment the potential gains from trade are siphoned away from low-income economies by structural features in the operation of multinational firms and international commodity markets (Singer, 1950;

Myrdal, 1957; and Prebisch, 1963; for a review of the origins of these doctrines, see Singer [1984]).

The first two criticisms rely on linkage theory (Hirschman, 1958). According to this approach, the development problem is not principally one of finding the best combinations of given resources, but of finding a stimulus that will call forth unutilized or poorly utilized domestic resources. Whether a leading mineral-export sector can perform this role will depend on its production, consumption, and fiscal linkages with other sectors.

Production linkages represent the potential for generating supplying industries (backward linkage) or processing industries (forward linkage). In high-rent activities the potential for production linkages will be low compared to other industries with similar total output value because the proportion of final value attributable to intermediate inputs and reproducible factors of production is lower than in industries with fewer or no rents.

Consumption linkages refer both to the potential for the domestic consumption of a sector's output and to the repercussion effects of leading-sector activity on the propensity of domestic consumers to purchase domestic goods and services in general. If the leading sector's activity tends to raise the propensity to import, or if its technological bias and the presence of foreign capital and management tend to suppress "learning by doing" effects in local industries, then consumption-linkage effects will be detrimental. In any event, the potential for local consumption of mineral-sector output remains minimal in most low-income countries until the conditions for resource-based industrialization are present (see Auty [1989]).

Provided the taxation problem has been addressed, the fiscal linkage (the contribution to government revenue and thus to public expenditure programs) from a mineral-export sector is likely to be the most important positive linkage. Emphasis on overall retained value—the proportion of export proceeds retained in the exporting country—can obscure the extent to which government revenue is its principal component. But the potential fiscal linkage can yield a disappointing outcome if the domestic savings rate is lowered, whether mineral rents are applied to the demand-side option (increase of consumption by transfers or subsidies) or to the supply-side option (public investment in infrastructure or directly productive activities).

Governments that increase investment spending in response to a minerals boom are likely to exhaust their range of high-yielding, quick-implementation projects in a very short time and to face domestic capacity constraints in construction, transportation, engineering, and various service industries (Garnaut and Clunies Ross, 1983; Lewis, 1984; Roemer, 1985; and Gelb and Associates, 1988).

Fiscal discipline in both expenditure and revenue raising often becomes an early casualty during a boom. Large, lumpy infrastructure projects that make heavy demands on future recurrent spending tend to be chosen for development. Rapid increases in public investment and consumption make wage restraints difficult to sustain, and the entry of highly paid, highly skilled foreign workers adds to the pressure to increase wages. Government and state corporations trying to invest in directly productive activities are no more likely, and perhaps less likely, to select economically viable projects than private investors. And the stimulus to growth from infrastructure built in advance of demand from industry and consumers is likely to take years to occur and to be limited in magnitude.

Nor is direct transfer of resources to the private sector, whether by reducing taxes or offering subsidies, necessarily an effective use of mineral revenues. Much depends on the regulatory environment and the structure of incentives facing private business. For example, there are often quick and very large returns to be made in real estate. Above all, a concentration of private effort on securing a share of windfall rents directly from government may be much more rewarding than investment in productive capacity (Krueger, 1974).

Once undertaken, spending obligations are difficult to reduce. Alternative industrial and agricultural production may have been weakened both by negative consumption linkages, such as diversion of consumption expenditure toward alternative imports, and by allocation decisions by government in favor of investments and consumption in other sectors or of imports. Public-sector savings performance may deteriorate very quickly indeed. The temptation to the government to resort to borrowing from the central bank has proved widely irresistible. Foreign exchange reserves then fall more rapidly still, forcing the government to resort to trade restrictions and to exchange and price controls.

Access to benefits from increased public expenditure and competition for a share of rents easily become the central political questions in a minerals boom. A minerals boom can be viewed as a special case of a mechanism by which "urban bias"—a systematic discrimination in pricing and resource allocation in favor of urban interests—is installed (Lipton, 1977). Profits, wages, and employment opportunities in urban services expand faster than their counterparts in rural production. Agriculture lags behind and receives low priority in public investment; it is less able than urban manufacturing to achieve sheltered status through protection. Capacity constraints and wage pressures increase the costs of public sector, industrial, and other service activities while budget laxity allows civil service posts to proliferate. Very large scarcity rents become available to those with control over resources such as housing, land, and transport facilities; rents also become available to those able to clear obstacles to obtaining the contracts or licenses necessary for participation in the construction and services boom.

There may be powerful reasons why governments—ostensibly representing the collective interest—fail to prevent these developments. Policies that introduce market distortions that cause overall welfare loss and that, in general, appear economically irrational may make a great deal of political sense if the state is viewed "as an agency for aggregating private demands . . . and public policies [are viewed] as choices made in response to political pressures exerted by organized interests" (Bates, 1983, p. 121), and if, furthermore, the government that operates the "agency" of the state has an interest in keeping itself in power. Clearly, coping with a minerals boom (or slump) is not only a matter of seeking the best policy package from the point of view of economic efficiency. It also requires a coalition of political forces that perceives a strong interest in avoiding the potential negative consequences of booming revenues.

The third structuralist criticism, that gains from trade are siphoned off from low-income economies to foreign investors, merits brief attention, even though this paper focuses on the domestic rather than the international dimensions of mineral rents. The view that the potential gains from mineral exports are eroded by remittance of profits overseas, reliance on imports of capital and intermediate goods

(and of skills), and location of downstream processing in industrial consumer countries amounts to a statement that linkage effects are negligible. But, if the taxation problem is soluble, remittance of profits is not
by itself detrimental when it represents a reasonable return to capital
and permits the maximization of retained rent for the resource-owning
country. Similarly, if imported capital supplements domestic savings and
imported goods permit comparative advantage to be exploited, the
positive or negative effects of the mineral industry will depend on the
way the fiscal linkage is forged, not on the fact of importation itself.

The other mechanism by which, it was argued, the financial
benefits of mineral exports might be siphoned away from resource-
owning countries is through a secular deterioration in the terms of trade
for primary commodity exports in relation to imports of manufactures.
This thesis began as a comparison of trends in international market
prices for primary commodities versus manufactures. The hypothesized
deterioration rests on three sets of market and institutional characteristics. First, it was observed that primary commodities face a very low
income elasticity of demand in their main markets compared to manufactures. Second, it was argued that because of competition from other
producers of primary commodities, excess labor supply, and protection
of primary commodity production in consuming countries, producers of
primary commodities exhibit an inability to appropriate the benefits of
productivity increases to wages and profits, the benefits passing
instead to processing industries or to final consumers. Third, it was
suggested that exporting countries cannot easily shift labor, capital, and
other inputs out of the production of primary commodities in the face of
declining terms of trade because of their small and fragmented domestic markets, the disproportionate availability of factors of production,
and foreign ownership.

The impact of these structural effects in diminishing benefits to
commodity-exporting countries will vary widely with the type of commodity, especially with the organization of its market and the range of
techniques available for its production, and with the resource endowment of the country. If the production function for the product is closely
related to factor availability in the producing country, and if there is a
wide range of actual or potential input–output relations, then benefits
through linkage effects will be greater; similarly, consumption linkages
will depend on the type of market that is created internally by the alloca-

tion of wages and profits. Arguments based on the variability in factor proportions and income distribution have been used to explain the high income levels attained by primary commodity exporters, such as Canada and Australia, in contrast to those achieved by today's low-income commodity exporters.

The trend in the mining industry has until recently been toward increasing capital intensity, scale of operations, and skill intensity. This trend has probably been assisted by the tendency of international mining companies to practice uniform, rather than locally specific, policies on choice of technique. Despite the availability of apparently cheaper unskilled labor in low-income economies, investment has proceeded on the assumption either that labor costs will rise in the future or that the benefits of low wages do not offset the disadvantages of lack of industrial experience and, perhaps, high turnover. Accordingly, with the important exception of tin mining, choice of technique in major mines has been relatively inflexible.

These structuralist arguments and their use of linkage theory suggest that mineral-export growth is likely to be a curse, not the blessing presumed by orthodox theory. Indeed, the principal political use of the structuralist case was to justify a switch away from development strategies based on primary exports and toward those based on import-substituting industrialization. Both orthodox and structuralist cases, however, suggest an inevitability of outcomes that discounts the room for maneuver of governments in devising solutions to the problems of securing and deploying mineral rents.

The Open Economy and the Dutch Disease

Modern macroeconomic analysis of the open trading economy, much used in recent years in the analysis of oil booms, offers a path through the conflicting cases just presented. The term *Dutch Disease* refers to the boom-induced rise in the real exchange rate and the associated relative decline in nonmineral traded-goods industries. It was coined to describe the effects on the Netherlands economy of the offshore gas discoveries in the late 1960s. The more neutral French term *Syndrome Hollandais* leaves open the question of whether the phenomenon is a "disease" and whether special treatment is indicated (Van Wijnbergen, 1984). Adjustment to the real exchange rate is necessary if an economy is to absorb its new mineral revenues; the difficult issue is whether measures can or should be taken during a boom to offset possible difficulties in a subsequent slump.

The symptoms of boom conditions in a mineral-export sector are as follows: In the construction phase of projects, there are large inflows of foreign capital and a substantial diversion of real resources to the new export sector. Wages and the prices of nontradable goods (for example, urban land or housing) are bid up, and a higher domestic general price level tends to result. The competitive position of those other sectors that produce tradable goods deteriorates.

The process is continued in the production phase if mineral exports are profitable. The relatively capital-intensive mineral-export sector will be in a weak position to resist wage demands, and higher real wages may spread to other sectors where productivity conditions do not warrant them. The substantial inflow of foreign exchange will tend to raise the price of domestic currency under a flexible exchange rate regime or produce more rapid inflation under a fixed-rate system. In either case the competitive position of preexisting export industries and import-competing industries is likely to be eroded.

If the inflow of foreign exchange directly raises the domestic money supply (that is, is spent by government or, if deposited in the domestic banking system, is lent to the private sector), then the money supply will accommodate further inflationary pressures. Where mineral revenues normally accrue to the state rather than the domestic private sector, the likely first-round expenditures are either on large capital projects that can be readily implemented or on the expansion of public services. In both cases cost pressure is intensified, and there is a relative contraction of traded-goods production outside the mineral-export sector.

When mineral revenues are diminished, other export industries and import-substitution industries are in a weakened condition that makes them unable to make up for the loss of foreign exchange, while the expansion of the public sector leaves the state with numerous commitments increasingly difficult to fulfill. There will be strong pressure to maintain domestic living standards for as long as possible by expanding the budget deficit and the money supply, thus worsening an already deteriorating balance-of-payments position.

The framework used to analyze the Dutch Disease is the "dependent economy" model of an open trading economy and its macroeconomic policy problems, built on the work of Meade (1951) and Salter (1959).[2] The model is recognizable as a "specific factors" model of a trading economy; the economy is dependent in that its import and export prices are fixed on the international market.[3]

There are two basic versions of the analysis. In the first, the economy consists of two sectors, traded and nontraded goods, and the minerals boom is treated as a windfall increase in the economy's ability to import traded goods. This version permits a focus on the demand side, or spending effects of a boom, and is useful where the switch of scarce resources into minerals production is not a major issue, as is often the case in low-income economies. In the second version, a three-sector analysis, a "booming" minerals sector with its own pattern of resource use is introduced. This permits interlocking examination of "spending" (demand-side) and "resource-movement" (supply-side) effects.

The three-sector analysis initially assumes that there are no idle domestic resources and that the external accounts are in balance. The spending effect of a boom (the overall rise in public and private expenditure permitted by the windfall foreign exchange gain) raises the prices of goods and services that can be supplied only from domestic resources (nontraded), whereas increased demand for output from the lagging traded sector can be met by imports whose prices are internationally determined. This "real" appreciation of the exchange rate (the price of nontraded relative to traded goods) causes a flow of labor out of agriculture and manufacturing (traded sectors) and a corresponding

---

[2]The adaptation of the model to the problems of mineral booms can be traced from Gregory (1976) through Snape (1977) to a number of papers appearing in the early 1980s; an explicit account of the full model was first presented by Corden and Neary (1982) and refined by Neary and Van Wijnbergen (1986).

[3]In a "specific factors" model, one or more factors of production (for example, capital or a natural resource) are specific to one sector and immobile between sectors. The basic Dutch Disease analysis usually assumes specific and immobile capital and mobile labor. The "dependent economy" assumption also means that relative prices of imports and exports (the terms of trade) cannot be altered by domestic policy action; hence, imports and exports can be treated as a single, composite "traded" commodity.

reduction in output in those sectors. The extent of the effect depends on the propensity to consume nontraded output; in mineral-exporting economies where increased government spending on construction and public services is likely to be the main channel for use of mineral rents, the marginal propensity will be high.

The supply-side, or resource-movement, effect occurs because the marginal productivity of labor in the booming sector has risen relative to that in other sectors. A withdrawal of workers from service industries (the nontraded sector) causes prices to rise at any given level of demand, while imports again fill the gap in other traded-goods sectors. The result is an effect similar to that of the spending (demand-side) effect. The strength of the supply-side effect depends on the relative use in each sector of labor and other productive resources: Where the booming mineral sector uses relatively little mobile labor (as is likely), the resource-movement effect may be dwarfed by the spending effect.

For many poorer mineral exporters, an assumption of unskilled labor mobility and immobility of capital (including a specific factor such as skilled labor in the mining industry) is probably realistic. In the long run, different assumptions about factor mobility become necessary and produce different results: Where capital can be shifted between the lagging traded sector and the nontraded sector, the outcome depends on the relative factor intensity of the two sectors.

Given the assumptions, a minerals boom brings about a rise in real income, a contraction of the lagging sector, and probably an expansion in nontraded goods production—hence, the possibility of deindustrialization in economies such as those of the United Kingdom, the Netherlands, and Norway and of a squeeze on agriculture in low-income countries. Wage–price flexibility and labor mobility ensure there is no unemployment. When the boom subsides (or collapses), the process should then reverse itself—again without unemployment (since wages are presumed to fall to market-clearing levels and the real exchange rate to depreciate). The level and future path of real income depend on the allocation of the windfall between consumption and investment and the efficiency of the investment undertaken.

The Dutch Disease approach is valuable because it highlights the importance of two key variables: the real exchange rate and the real wage. In addition, the assumptions of the basic model can be relaxed to cater to the circumstances of individual countries. The important assumptions are the initial position of internal and external balance, flexibility of wages and prices, the smooth adjustment process from one production structure to another, the absence of market distortions, and the "specific factors" assumption. The structuralist and linkage approaches can be viewed as a critique of these assumptions.

Economic Performance in Mineral-Exporting Countries

The discussion so far leads to the question of whether the economic characteristics of mineral exports and the processes by which mineral revenues are transmitted through the economy might lead to common characteristics among mineral-exporting countries and a shared performance record.

A Special Case?

The view has been advanced that mineral-exporting countries differ structurally from other low-income countries (including countries exporting agricultural primary products) and that their development performance has fallen behind countries with similar initial income levels but with a different composition of production and trade (Nankani, 1979). For sub-Saharan Africa, Wheeler (1984, pp. 8–9) suggests that the share of nonoil minerals in total exports immediately after independence is an important factor in explaining poor economic performance.

Mineral-exporting countries, however, are a diverse group. Among those for which fuels, metals, and minerals accounted for more than a quarter of exports in recent years are a few low-income countries (World Bank classifications)—Zaire, the Central African Republic, Sierra Leone, and Guinea—and numerous countries whose lower-middle-income status is often attributable more to mineral wealth than to broadly based income generation. These include the other African exporters of hard rock and precious minerals (Mauritania, Liberia, Zambia, Zimbabwe, Botswana); a number of "high-absorbing" oil producers with low incomes outside the oil sector (Indonesia, Nigeria, Congo, Ecuador); and a varied group of more diversified economies (Bolivia, Morocco, Papua New Guinea, Cameroon, Peru, Tunisia). The upper-middle-income group includes the diversified economies of Chile, Malaysia, and Mexico and five important oil exporters. The "capital surplus" oil exporters of the Gulf remain. World Bank data exclude centrally planned and nonmember economies.

Nankani (1979) compared a set of fuel and nonfuel mineral exporters with a control group of middle-income nonmineral economies. Although he found no clear performance patterns within the mineral exporters' group, there was evidence of poorer performance for this group than for the control group in a number of respects. Nonfuel mineral economies had lower incremental savings rates (although oil economies did not). The mineral economies experienced greater technological dualism (extremes of capital or labor intensity in production), wider intersectoral wage differentials, higher unemployment, and lower school-enrollment ratios. Inflation rates tended to be higher in mineral economies. Agriculture tended to grow more slowly, and food constituted a larger share of total imports. Nankani suggested that mineral economies were in fact more prone to export earnings instability than were nonmineral economies and that their exports tended to remain more concentrated.

There are considerable difficulties in arguing the thesis that mineral-exporting countries share a common record of overall economic performance. Statistical analysis of the kind cited leaves out of account the state of a country prior to the advent of a mineral sector and the continuing role of the preexisting nonmineral sectors, something to which Dutch Disease economics usefully points. Furthermore, there may be very wide disparities in national average income per capita in different sectors of the economy, and little to distinguish the nonmineral sectors from those of other countries without a mineral-export sector but with similar population, ecological, or historical characteristics. Thus, a comparison of selected indicators for mineral-exporting countries with a control group of nonmineral exporters selected by income level alone may not be a valid comparison at all, because ceteris paribus conditions are most unlikely to hold.

As illustrated in the next section, there actually are wide disparities of performance among mineral exporters. But there are no convenient measures of the political factors constraining governments in tackling the economic challenges that arise during the growth, continuation, or decline of a mineral-export sector.

Table 1. Minerals as a Percentage of Total Exports and Per Capita Income for the Less Developed Countries Exporting Nonoil Mineral Commodities

Country	Minerals as % of total exports		Per capita income (U.S.$)
	1965	1985	1985
Africa			
Zaire	72	74	170
Togo	33	52	230
Sierra Leone	25	34	350
Ghana	13	30	380
Zambia	97	94	390
Mauritania	94	58	420
Liberia	72	65	470
Morocco	40	32	560
Zimbabwe	24	25	680
Botswana	—	76	840
Caribbean and Latin America			
Bolivia	93	82	470
Jamaica	28	67	940
Peru	45	70	1,010
Chile	89	64	1,430
Asia-Pacific			
Papua New Guinea	—	51	680
All low-income countries[a]	24	23	200
All lower-middle-income countries	28	51	820

Source: Adapted from World Bank, 1987, *World Development Report: 1987,* Washington, D.C.

[a]World Bank classifications, excluding China and India.

Experience of Nonoil Mineral-Exporting Countries

Over the period 1965 to 1985, the average low-income country (World Bank classifications, excluding China and India) had slow growth of per capita income, with the worst performance concentrated in the first half of the 1980s. Lower-middle-income countries fared considerably better. The mineral exporters included in tables 1 and 2 had overall growth rates widely dispersed around the average, but in general these were strongly influenced by the expansion or relative contraction of their mining sectors.

Among low-income countries in general, sectoral growth of value added (outside the mining sector) was most rapid in manufacturing,

99

Table 2. Development Indicators for Less Developed Countries Exporting Nonoil Mineral Commodities, 1965 to 1985

Country	Annual growth of per capita GNP (%) 1965–1985	Share of GDP (%)				Total government expenditure as % of GNP		Average annual inflation rate (%)	
		Gross domestic investment		Government consumption					
		1965	1985	1965	1985	1972	1985	1965–1980	1980–1985
Low-income									
Zaire	-2.1	14	13	10	6	19.8	23.3	24.5	55.3
Togo	0.3	22	26	8	14	—	42.0	7.1	6.9
Sierra Leone	1.1	12	9	8	12	—	15.4	7.8	25.0
Ghana	-2.2	13	9	14	9	19.5	12.5	22.8	57.0
Zambia	-1.6	25	12	15	19	34.0	30.3	6.4	14.7
All low-income countries[a]	0.4	15	15	11	12	18.0	20.3	11.4	18.9
Lower-middle-income									
Mauritania	0.1	14	25	19	15	—	—	7.5	8.1
Bolivia	-0.2	22	17	9	9	9.6	39.9	15.7	569.1
Liberia	-1.4	17	9	12	21	—	28.2	6.5	1.6
Morocco	2.2	10	22	12	16	22.8	33.5	5.8	7.8
Zimbabwe	1.6	15	23	12	19	—	39.1	5.7	13.2
Papua New Guinea	0.4	22	22	34	23	—	35.8	8.1	5.5
Botswana	8.3	6	21	24	23	33.7	48.2	8.0	5.2
Jamaica	-0.7	27	23	8	16	—	—	12.6	18.3
Peru	0.2	15	20	10	11	16.7	12.9	20.5	98.6
Chile	-0.2	15	14	11	14	43.2	35.5	129.9	3.9
All lower-middle-income countries	2.6	18	20	11	13	19.4	24.8	22.2	22.3

Note: GNP = gross national product; GDP = gross domestic product; dash = data are not available.

Sources: Adapted from World Bank. *Economic Data,* vol. 1 of *World Tables,* 3rd ed., 1983; *World Development Report: 1987.* Washington. D.C., 1987.

[a]World Bank classifications, excluding China and India.

followed by services, and slowest in agriculture. This pattern applied in both the 1965–1980 and 1980–1985 periods. The same pattern, at higher rates overall, applied to the lower-middle-income countries.

Individual mineral-exporting countries, however, exhibited markedly different patterns. Of the low-income group, only Zambia and Sierra Leone experienced a rapid expansion of manufacturing in the 1965–1980 period. The expansion in both cases was concentrated in the early part of the period when still-buoyant mineral revenues permitted a large increase in demand for the output of protected manufacturing. For Zambia the effect was compounded by regional circumstances: Zambian industry expanded at a phenomenal rate once the Zambian market was closed by sanctions to the output of the Rhodesian (Zimbabwean) manufacturing industry.

Togo and Sierra Leone both experienced the increase in relative importance of services that the Dutch Disease model predicts. For Togo the cause was a large increase in phosphate revenues in the early 1970s. The Sierra Leone experience is severely complicated by the prevalence of smuggling, and the country's official export purchasing power is recorded as declining throughout the 1970–1985 period (table 3).

Although Zaire remained a substantial mineral exporter of copper and cobalt throughout the 1965–1980 period, its export purchasing power consistently declined. The relative share of GDP of both manufacturing and services declined and, particularly toward the end of the period, agricultural output increased. Assuming most manufacturing was either sheltered (nontraded in character) or closely linked to the mining industry, the case of Zaire contains some elements of the reversal of Dutch Disease. But it also shows that throughout the life cycle of a large mining sector, development patterns form that are quite different from those experienced in a "typical" country at a similar level of income.

The portion of minerals in Ghana's exports remained low compared with the other countries in the sample. Ghana was also a locus classicus of economic deterioration brought about by political upheaval and the pursuit of highly distorted exchange, trade, and budget policies during the 1960s and 1970s. For these reasons it is not of great help in understanding the economic impact of a mineral sector. In the 1980s, however, the expansion of its mineral exports (largely gold) resumed as

Table 3. Changes in Export Purchasing Power and Real Exchange Rates for the Principal Less Developed Countries Exporting Nonoil Mineral Commodities, 1970 to 1985

Country	Change in export purchasing power[a] (%)			Real exchange rate index[b] (1980 = 100)			
	1970–1975	1975–1980/81	1980–1985	1970	1975	1980	1985
Zaire	−18	−8	−11	66	70	100	39
Togo	43	−5	−12	54	100	100	74
Sierra Leone	−9	−10	−5	159	102	100	196
Ghana	−10	−10	−4	34	29	100	53
Zambia	−21	−12	−11	142	102	100	84
Mauritania	−14	−16	39	119	110	100	112
Bolivia	12	15	−4	137	95	100	112
Liberia	−20	−26	−9	143	109	100	134
Morocco	8	−6	2	126	103	100	71
Zimbabwe	—	—	—	160	118	100	91
Papua New Guinea	—	17	1	120[c]	105	100	105
Botswana	—	—	—	—	102	100	79
Jamaica	9	−9	−19	149	122	100	79
Peru	−2	−4	−3	191	158	100	102
Chile	1	7	−2	—	88	100	72

Note: Dash = data are not available.

Sources: Calculated by author from World Bank, Country Economic Data Sheets 1 and 2, in Economic Data, vol. 1 of World Tables, 3rd ed., 1983; World Bank, World Development Report: 1987, Washington, D.C., 1987; International Monetary Fund, International Financial Statistics, Washington D.C., December 1987.

[a]Cumulative result of annual average or period absolute change in terms of trade and export volume applied to base-year (first of period) nominal exports and expressed as a percentage of base-year gross domestic product.

[b]Nominal index of dollars per local currency unit multiplied by local consumer price index and divided by nonoil less-developed-country import price index. An increase from one year to the next indicates an appreciation of the real exchange rate during that period.

[c]1971.

part of a thorough program of economic reform. If Dutch Disease symptoms can be avoided, Ghana may in the future offer useful lessons in economic management.

The mineral-exporting countries in the lower-middle-income group exhibit a greater variety of experience. At least seven of the sample experienced a mineral boom at some time during the 1965–1985 period: Mauritania, Bolivia, Morocco, Papua New Guinea, Botswana, Jamaica, and Chile. The boom usually was due to a substantial expansion of mineral output, but price booms, particularly for copper, tin, and precious metals, have also played a part. In most of these cases, the pattern of development has more closely matched the predictions of the Dutch Disease model.

Bolivia, Morocco, and Botswana had relatively rapid growth of services, and Jamaica's manufacturing sector sharply contracted in relative terms. Botswana's principal preexisting traded sector (agriculture) also contracted in relative terms, although drought, rather than the minerals boom itself, appears to have been the major cause.

The effect of the country's stage of development at the beginning of the period, and of policies pursued, is exhibited by the contrast between Chile and Papua New Guinea. Chile saw a relative contraction (and, for a time, an absolute reduction) in manufacturing output from a mature base; exchange rate and trade-liberalization policies bolstered by both foreign borrowing and mineral revenues had a devastating effect on previously sheltered domestic manufacturing. In Papua New Guinea the lack of manufacturing industry at the beginning of the period and the potential for its expansion on the basis of agricultural export processing and unsatisfied domestic demand meant that positive demand-side effects of increased revenues outweighed potential negative supply-side effects. Papua New Guinea exported a relatively wide range of primary commodities during the period and was thus somewhat protected against large fluctuations in earnings from any one commodity; the stabilization policies pursued also helped to prevent an appreciation of any magnitude in the real exchange rate over the 1975–1985 period.

Countries with declining mineral export purchasing power in the lower-middle-income group also include cases where the reversal of Dutch Disease is apparent. Liberia, Peru, and even Chile (1980 to 1985) showed relative contraction of services in the process of adjustment to declining mineral-export revenues.

In all cases, other factors are at work: other exports, foreign borrowing, and aid flows. Each of these can bring about effects similar to those caused by mineral revenues, and their increase can offset trends that would otherwise arise from a decline in mineral-export earnings. Nevertheless, the broad outlines of the statistical trends confirm the expectations of the Dutch Disease model.

A simple analysis of trends in real exchange rates is presented in table 3. The real exchange rate index used is the conventional calculation of the index of foreign currency units per local currency unit multiplied by the local consumer price index and divided by the nonoil less-developed-country import price index.[4] Specifically, a bilateral real exchange rate with the U.S. dollar is used, along with a dollar-denominated international price index.[5]

The exchange rate data in table 3 yield a more ambiguous picture than trends in the structure of production. During booms in export earnings, sharp appreciation of the real exchange rate was experienced by Togo (1970 to 1975), Mauritania (1980 to 1985), and Chile (1975 to 1980). Declines in the real exchange rate are associated with drops in export income in Togo (1980 to 1985), Zambia (1970 to 1985), Jamaica (1975 to 1985) and, less clearly, Peru (1970 to 1980). In a number of cases, appreciation of the real exchange rate accompanied a decline in export income: Zaire (1970 to 1980), Sierra Leone (1980 to 1985), Ghana (1975 to 1980), Bolivia (1980 to 1985), and Liberia (1980 to 1985). Depreciation of the real exchange rate went hand in hand with an export boom in Bolivia (1970 to 1975), Morocco (1970 to 1975), Papua New Guinea (1975 to 1980), Botswana (1980 to 1985), and Jamaica (1970 to 1975). In other cases there was either little significant association or relative exchange rate stability.

These "perverse" results do not necessarily invalidate the Dutch Disease model. Continuing appreciation of the exchange rate when export purchasing power is declining commonly results from the impo-

---

[4]The calculation is conventional in that a comparable statistical routine is used by the International Monetary Fund in making such calculations. Thus, the formula does not match the theoretical formulation of the real exchange rate as the relative price of nontraded goods in terms of traded goods in the domestic market. There is a mathematical relationship between the two concepts, governed by the structure of output (relative weight of traded and nontraded goods) and divergencies between the domestic and foreign prices of traded goods. If trade policy changes, the two measures may diverge; nevertheless, the conventional statistical formulation is adequate for present purposes and is, in any case, the only one feasible to compute without relative price models for each economy.

[5]This procedure avoids the need to choose a basket of foreign currencies against which to measure the local currency's value. Some sophistication is lost, and the conclusions are vulnerable to the sharp appreciation of the U.S. dollar in the 1980s. The bilateral rate procedure follows that used by Roemer (1985) in his comparable analysis of development trends in oil-exporting countries.

sition of exchange and trade controls in conjunction with a widening budget deficit. In other words, the government prevents the automatic adjustment of the exchange rate from taking place and appropriates an increasing proportion of available foreign exchange to public-sector uses at an artificially low price in terms of local currency. The exchange rate measured is the official rate and therefore does not include the scarcity premium that would be attached in parallel market transactions. This explanation applies in some degree to all the perverse cases of appreciation noted: Zaire, Sierra Leone, Ghana, Bolivia, and even Liberia (where, in the 1980s, the formal currency of the U.S. dollar previously in use all but disappeared in favor of a locally printed nonconvertible substitute).

Depreciation in the face of an export boom has less uniform explanations. For Bolivia, Morocco, and Jamaica, it reflects an overdue adjustment to preexisting conditions of disequilibrium that were not offset by increments to mineral revenues. In Botswana and Papua New Guinea, balance-of-payments equilibrium was maintained throughout by appropriate stabilization measures leading to the accumulation or decumulation of foreign assets according to the short-term state of export revenues. Absorption (consumption and domestic investment) was not permitted to adjust automatically to the availability of income, and thus an accommodating change in the real exchange rate was avoided in favor of broad stability against a basket of currencies. Real depreciation against the U.S. dollar resulted partly from successful containment of domestic cost increases and partly from the dominance in each country's currency basket of one relatively insignificant currency in terms of international trade: the South African rand in Botswana and the Australian dollar in Papua New Guinea. Both currencies fluctuated sharply against the U.S. dollar, but around a general trend of relative depreciation.

Even those cases where the direction of exchange rate changes was more in line with expectations do not necessarily exhibit the extent of change in the exchange rate that would have occurred in the absence of exchange and trade controls or countervailing absorption adjustments enforced by deliberate economic policy action. In Zambia, Zimbabwe, and Jamaica, for example, the extent of real depreciation was far from sufficient for exchange controls to be removed; in each case, a substantial public-sector deficit persisted through the period of falling mineral income.

These results bring into focus an important property of the Dutch Disease analysis. In common with other economic models, it traces the repercussions of specific changes in variables in a given context (initial equilibrium, price flexibility, and so on) and with the ceteris paribus assumption applying to variables not altered independently or within the logic of the model. The Dutch Disease model appears to be robust in a wide variety of cases, but deliberate government policy can reverse the predicted outcome; the policy effect can be incorporated in a more elaborate version of the model.

Thus, the pathology of the "disease" is not independent of the economic policy response of government. Government can either exacerbate or mitigate the effects that would occur in the absence of intervention; it can prevent some of the symptoms in a boom and it can prolong them in a slump. A similar conclusion emerges from the experience of oil exporters (Gelb and Associates, 1988).

Economic Policy Choices

The governments of mineral-exporting countries face three important policy issues: the taxation problem, the macroeconomic problem, and the absorption problem.

The Taxation Problem

To avoid distortion of exploration, development, and production decisions in the minerals industry, the reproducible factors of production and intermediate inputs required have to be paid for at going market rates. In the case of capital, investors must be able to expect returns at least at market rates, net of government impositions, if a project is to go ahead. This criterion requires that a taxation system disturb as little as possible that portion of expected pretax returns that represents the necessary reward to capital. To achieve this and to maximize government revenue both from an individual project at a given tax rate and from a succession of projects that are feasible before tax, a taxation system must aim at neutrality with respect to the choice of technique and pace of extraction in mineral projects. In short, an efficient and neutral mineral tax system will use as its tax base the Ricardian rent in mineral revenues. The proportion of pure rent that can be taxed away will depend, in part, on the taxation regimes of other countries with potentially competing mineral deposits.

In principle, the taxation problem is avoided if there is state ownership of the mining industry, as well as the resource in the ground, and if the state is able to supply the necessary capital. Even in this extreme case, however, the taxation problem is likely to reappear as a choice about the dividend policy to be imposed on the state enterprise responsible for mining activity. In most cases in the real world, foreign or private capital and skills are required to some degree even for state-owned enterprises; their reward is once again part of the taxation problem in its broad sense. The international trend appears to have turned once again toward foreign and private ownership of minerals industries, or at least toward forms of state ownership and control that amount, in economic content, to the same thing (such as production-sharing contracts between state companies and private investors in the petroleum industry). The taxation problem is therefore presented, in one form or another, to all governments in mineral-exporting countries.

The magnitude of Ricardian rent from prospective mineral deposits cannot be determined in advance. Knowledge about the likely value of a deposit will shift from the mining company to the government as exploration, development, and production proceeds. If, however, a government proposes to rely on this shift and impose variations in taxes designed to secure most of the rent, as project revenues are realized there will be a substantial increase in *ex ante* risk to the investor. As compensation for this risk, the supply price of capital (the required return) will rise; economic rent available to the government will thereby be reduced, both because the cost of capital is higher and because some deposits will not be developed.

A possible solution is to impose a system of taxation, fixed *ex ante*, which includes a proportional (and relatively high) rate of tax on discounted cash flow returns in excess of a predetermined percentage rate that represents an appropriate return on capital invested. Systems of this kind are now in place in a number of countries for petroleum and nonfuel minerals, including Australia, Papua New Guinea, Seychelles, Mozambique, Tanzania, Equatorial Guinea, Ghana, and Sierra Leone. Major producing countries such as the United Kingdom, Indonesia,

Angola, and Botswana have different systems, but they approximate the same effect.

The principal problem with such systems is determining the appropriate *ex ante* rate of return to capital (see Garnaut and Clunies Ross [1983]). Where there is competition for a deposit, the rate can be discovered through a bidding process. Where a "bilateral monopoly" exists between a single potential investor and the government, a process of negotiation relying on knowledge of going rates in the international market is necessary. One solution, adopted with increasing frequency, is to set the tax at progressively higher total rates in bands of income that represent increasing threshold rates of return; in this way, the risk of misspecifying the rate is minimized by using a band around the probable target.

It is unlikely that all potential rent will be captured by this form of tax when, as is usual, it is imposed in conjunction with other forms of normal business tax (administrative considerations, for example, may require that the threshold return be the same across all deposits, whereas efficiency considerations require that some portion of rent remains with the taxpayer). But the system is likely to come closer to this objective of using rent as the tax base than would sole reliance on the customary royalties or income taxes.

Note, however, that a system that efficiently taxes mineral rent may maximize revenue flows at the "cost" of increasing their instability. Part of the macroeconomic problem, therefore, is to determine whether it is possible to manage aggregate demand in ways that will minimize the damage from such instability.

The Macroeconomic Problem

The extremes of possible responses to shocks can be summarized using simple national account identities. If income equals expenditure, then

$$Y = C + I + G + X - M$$
$$= A + B$$

where Y is national income; A is total domestic expenditure (absorption) consisting of private consumption (C), investment (I), and government consumption expenditure (G); and B is the trade balance of exports (X) minus imports (M). If some shock lowers national income,

Y, then either absorption, A, or the trade surplus, B, must fall. In policy terms, either domestic expenditure must be cut or a trade deficit must be financed. An increase in national income (through a mineral boom, for example) presents the mirror image of these options: Either domestic expenditure must be increased, or foreign financial assets must be accumulated as the counterpart of an increased trade surplus.

If a slump or boom is temporary (for example, if the cause is a cyclical swing in the world economy), a "constant absorption" strategy (holding the rates of consumption and investment expenditure steady) is feasible in the case of a slump and desirable in the case of a boom. A trade deficit can be met by borrowing, on the principle of the International Monetary Fund's Compensatory Financing Facility, whereas a trade surplus can be used to accumulate foreign assets, the stock of which will be reduced in the cyclical downturn.

The alternative policy extreme is to seek to restore external balance. In a simple model, this would imply the condition that $B = 0$. With a positive shock, an increase in absorption sufficient to eliminate the trade surplus is easy to obtain; the reduction in absorption required to eliminate a trade deficit in the case of a negative shock is much harder to implement. This is known as the asymmetry of adjustment.

Moving from the simplified adjustment alternatives just presented and allowing for capital movements or alterations in savings, there are four broad paths of adjustment, each with differing impacts on growth and distribution. For simplicity, consider them as responses to a negative shock:

1. *Constant absorption* This requires additional real external financing.
2. *Trade adjustment* This entails raising export responsiveness to external demand, reducing import responsiveness to growth in gross domestic product (GDP), or both.
3. *Increased domestic savings* An excess of expenditure over income implies an excess of investment over saving. A rise in savings requires a reduction in private consumption relative to GDP, a reduction in the government deficit relative to GDP, or both.
4. *Reduced investment* A fall in investment relative to GDP affects external balance through the same mechanism as path 3.

Paths 3 and 4 are strongly contractionary, at least in the short term, and path 4 is also likely to constrain the long-run rate of growth.

Path 1 is not initially contractionary, but its sustainability is subject to the availability of external finance. Path 2 is not contractionary unless export expansion directly induces offsetting imports.

The basic Dutch Disease analysis illustrates how relative prices adjust in the face of a boom to restore external balance and how resources may move out of preexisting traded-goods production into the booming sector and possibly into nontraded activities. The impact will differ according to the elasticity of factor substitution between traded and nontraded activities; the flexibility of wages and prices; and the degree to which there are unemployed resources, especially labor. The higher the elasticity of substitution, the more flexible the prices and wages, and the greater the extent of unemployed resources, the lower the need for real exchange rate adjustments. The greater this "flexibility," the better the economy's ability to adjust to instability in mineral windfalls and the higher the potential rate of noninflationary growth in output outside the mineral sector.

An important task of macroeconomic management is therefore to frame a realistic estimate of the rate of medium-term capacity growth the economy can sustain without encountering strong inflationary pressure (with its counterpart in physical supply bottlenecks, industrial conflict, and the generation of quasi rents for groups such as real-estate owners or skilled workers). The pace of absorption (consumption and domestic investment) of the mineral windfall then has to be kept within this potential capacity growth rate; the portion that cannot be used is better saved abroad, to be utilized when the economy's range of potentially profitable activities has expanded sufficiently to allow a real return on additional investments greater than the yield on assets held abroad. In a study of six "high-absorbing" oil exporters, Gelb and Associates (1988) estimated that long-term economic performance would have been better than it actually was if as much as two-thirds of the 1973–1974 and 1979–1980 oil windfalls had been saved abroad.

If it were possible to forecast with certainty the availability of revenues, and if revenues were to flow without fluctuation, the estimation of a safe rate of capacity growth would be sufficient for a decision on how much to spend or save year by year. Uncertainty and fluctuation, however, must be taken into account.

Any forecast of revenues implies a probability distribution of outcomes. Maintaining a safety margin in the face of uncertainty requires selection of a forecast below the mean of the distribution to be chosen.⁶ In other words, a precautionary and conservative policy stance requires mineral price forecasts that are more likely to under-estimate than overestimate actual revenues.

Revenue fluctuations raise the problem of asymmetry of adjust-ment. Maintaining internal and external balance over the medium term is desirable for the reasons already discussed. In addition, it is much easier to raise expenditure in the upswing than it is to reduce it in the downswing. Once the medium-term path of expenditure growth has been chosen on criteria of absorptive capacity and the likelihood of errors in revenue prediction, the impact of short-term revenue cycles can be dampened by adopting a strategy of constant absorption. To do this, the government will either save (in foreign assets) during the upswing and draw on these reserves during the downswing, or borrow in the downswing and repay in the upswing. In either case the rate of change in public expenditure is relatively stable from year to year.

Both domestic and external pressures make the constant absorption strategy a difficult route to follow. In fact, they frequently obscure the likelihood that it is the optimal strategy. It involves treating external balance as a medium-term concept and tolerating wide swings in foreign assets in the short term. An image of short-term prosperity will be tempting to powerful interest groups within the country and perhaps cause external donors to view the country as less needy. An attempt to achieve constant absorption by countercyclical fiscal and monetary policy may fail because responsiveness is too low and fine-tuning is hazardous. In any case, frequent changes in tax provisions or interest rates may distort production decisions and defeat the long-run purpose of the strategy.

The strategy may be feasible where automatic stabilizers are available. This might be the case in countries with relatively unsophisti-cated financial markets: In boom times, foreign companies remit profits

---

⁶In two cases selection of a forecast with no more than a 20 to 30 percent chance of a lower actual price or revenue outcome than forecast has been recommended: Alaska limited budgetary expenditure plans out of oil revenue by statute to the 30 percent level (Gelb and Associates, 1988), and Garnaut and Baxter (1983) recommended for Papua New Guinea adoption of a gold-price forecasting rule for budget decisions in this 20 to 30 percent range.

overseas, thus removing demand from the economy. If export earnings fall, import capacity will fall at the same time; banks are likely to be aware of these characteristics and to seek to conserve liquidity in good times (thus dampening the growth of expenditure) to allow it to fall in bad times (thus allowing expenditure to be kept steady). In addition, institutional mechanisms can be designed to separate fluctuating mineral revenues from general government revenues and impose rules about the rate at which mineral revenues can be used for public investment and consumption (Papua New Guinea's Mineral Resources Stabilisation Fund is an example).

The noninflationary, precautionary, and stabilizing strategy outlined for the management of aggregate demand will also assist in "leaning against the wind" of boom-induced movements in the price of foreign exchange and the price of labor, the two key prices in the Dutch Disease analysis. Both the real exchange rate and the real wage tend to rise with national income growth in the long term; moreover, movements in these prices are a necessary part of the adjustment process to restore external and internal balance in response to a shock. But the pace and extent of these changes in a boom (or an unwarranted rise in what turns out to be a cyclical upswing) can reduce the economy's capacity to withstand a slump by inducing intersectoral resource shifts and changes in consumption patterns that are not sustainable and not quickly reversible. Some form of intervention—for example, to restrain wage rises in boom periods—may therefore be warranted.

The Absorption Problem

The single most important choice in the absorption strategy for mineral revenues is probably the one just addressed: How much of windfall revenue should be spent and how much should be saved abroad? An appropriate and cautious choice on the pace of absorption seems to be the key determinant of whether an economy will survive a slump and convert mineral revenues into an improved path of consumption and better overall economic performance in the long term.

Once this choice has been made, it must be decided how domestic investment and consumption of revenues will occur. If, as is usual, mineral revenues first accrue to the public treasury, then how much will go to public investment, how much will go to public consumption, and how much will be transferred to the private sector?

This allocation may not be entirely made by overt public choice through the processes of government. If there are strong trade unions, if there is oligopoly in nontraded service activities (such as transport or wholesaling and retailing), or if there is high concentration or some other restriction in land ownership, part of the mineral rent is likely to be appropriated privately as quasi rents in the form of higher wages, distribution markups, or land and property rents. The long-run effects of these involuntary distributions will depend on the propensity to save and invest of the groups that receive them. If the incentive to save and invest is adequate, the effects will not necessarily be negative, and growth engendered by private capital accumulation can follow.

If the environment is conducive to privately led growth, then deliberate distribution of mineral revenues to private agents will be worthwhile. But so far this strategy has been pursued mainly in advanced industrial countries with a highly developed industrial and infrastructural base already in place. Alaska and Alberta explicitly pursued this choice. This strategy also underlies the British government's approach in making direct tax reductions for firms and households (also financed partly by sales of public assets), although these reductions were largely offset by changes in capital allowances and indirect taxes—sustaining high unemployment through social security transfers and, for the longer term, reducing the stock of public debt and thereby reducing the potential tax obligations of future generations.

In most low-income countries, however, the path of public investment and consumption has been the predominant choice. Over recent decades the influence of structuralist and linkage theories of development has been strong. International conditions were considered too unfavorable, the range of market imperfections and failures too great, and the gaps between social and private rates of return too wide in key areas of investment to allow private decisions to prevail. The public sector was considered the right agent to undertake high-risk, large-scale investments as part of an overall program of social and economic transformation. Given the widespread record of failure with such approaches, and the particularly dismal experience of most oil exporters with the public investment of oil windfalls, the attitude of governments to private-sector initiative is changing. At present, few countries have mineral windfalls in sight, but, if they do recur, some change in the approach to absorption might be expected.

The choice to transfer revenues to the private sector is not independent of the choice between consumption and investment for the economy as a whole. Should the allocation be to firms or to households and in what form? A distribution to households carries the risk of creating an excessive consumption boom unless a high marginal propensity to save can be relied on. Virtually all forms of potential distribution to private firms, however, will disturb market signals governing investment decisions, thus returning the larger choices about the direction of investment to the government even if it is not itself engaged in productive activities.

Whether distributed to firms or households, a further choice exists between tax reduction, subsidies to particular activities or social and economic groupings, or subsidies on products. Tax reduction relies on the adequacy of the surrounding framework of incentives to induce efficient saving and investment decisions. Subsidies to groups will reflect an explicit view either about desirable income distribution or about the likelihood that these groups will invest in activities that the government deems most promising. Subsidies on products will be offered on a similar basis and will require a general strategy in favor of either promotion of exports or production for the home market.

The subsidy approach—widely used even where public investment has been the predominant choice—carries two risks: (1) that the subsidized products will require a permanent subsidy rather than one that can be progressively removed as mineral revenue is run down, and (2) that access to subsidy will become a dominant political issue.

The choice in public investment lies between directly productive activities and infrastructure. The favored productive activities usually lie in the nonmineral traded-goods sectors. As the linkage approach suggests, there often will be scope for promoting industries supplying goods to the mineral sector and for resource-based industrialization. The long-term usefulness of both is limited by the exhaustibility of the mineral resource itself, unless the capacity installed is efficient by world standards at postboom relative prices.

Import-substituting industrialization faces the obstacle of appreciation of the real exchange rate that accompanies the minerals boom; this tends to invite a protectionist approach and a general stance of inward orientation, such as that adopted for much of the post–World War II period in Latin America and later in sub-Saharan Africa, with generally poor results. Investment in agriculture ought to be favored by many low-income mineral exporters on grounds of comparative advantage, income distribution, and the likelihood that underemployed labor in the sector can be quickly mobilized. The barrier, however, is that public investment in agriculture has seldom been a success: The record of state farms (in Zambia, for example) has been uniformly bad.

For all these avenues, public-sector investment carries a number of well-known disadvantages. It is administratively convenient to choose large-scale projects (often larger than the potential market will sustain) and to construct on a turnkey basis using foreign contractors and unfamiliar imported technology. The future operation of such projects in the local environment may be vulnerable to lack of skills, technological advances elsewhere, and sudden changes in relative prices.

Rapid investment on a large scale brings rent seeking into the competition for contracts and increases the inflationary pressure on construction, transportation, and other nontraded services. A bias toward imported inputs is likely, which renders the plant vulnerable to a future foreign exchange deficiency when mineral revenues subside.

Investment in infrastructure amounts to investment in the hope of generating future traded activities. The same problems of rent seeking in the contract-awarding process and of bottlenecks in services arise. More important, however, is that in "normal" development, infrastructure needs are usually identified as a consequence of directly productive activities rather than in advance of them: The prevalence of "white elephant" roads, harbors, airports, and even conference halls in mineral-exporting countries is ample evidence. Most infrastructure projects yield no direct financial return—in foreign exchange or public revenue—to service debt (if raised in advance of revenues) or to provide replacement earnings if traded-goods activities are not induced.

The lesson from many of the newly industrialized countries of East Asia is that investment in social infrastructure—education and health—merits substantial emphasis. Such investment amounts to public creation of private human capital and at least offers a resource that can sustain itself after a boom. Nevertheless, these sectors carry high continuing recurrent costs, and, therefore, the ever-increasing social demand for education and health services may be difficult to meet when public revenues are reduced. The same, of course, is true of expenditure on improvement of general public administration, in itself the solution to a major bottleneck.

An important question in both the spending pattern and the general policy stance is whether existing traded-goods production should be protected in the face of a rise in the real exchange rate that threatens to put at least some businesses out of production. The theoretical answer, based on Dutch Disease analysis, is usually no, unless there are significant "learning by doing" effects that can be obtained by temporary subsidy during the period that mineral revenues are available (Van Wijnbergen, 1984). Appreciation of the real exchange rate and the associated industrial restructuring are necessary parts of the adjustment process an economy must undergo to enjoy its newfound wealth. Import liberalization, moreover, is one of the ways in which the adjustment can be made to happen with a lower rise than otherwise in the general price level. Nevertheless, just as there is a case for a precautionary rate of spending growth, there may be a case for precautionary support of activities with a likely comparative advantage (but substantial regeneration costs) after the boom. Tariff protection, subsidy, or deliberate action to retain rises in the nominal exchange rate are all available for this purpose, but carry considerable risk of introducing distortions, undesired distributional effects, or a higher rate of inflation.

Some Lessons for Economic Policy

Although there is a theoretical framework for analyzing price and quantity changes in the face of mineral booms, and although the developmental record of mineral-exporting countries is for the most part poor, there is little a priori inevitability about the outcome of mineral booms and slumps. It is possible to turn a mineral windfall to advantage, but it is also possible to create a development pattern that is worse (in terms of welfare) than that which would have been in place without the minerals. The outcomes depend largely on the things governments do, and thus on the pressures and interests that form and influence governments. In this concluding section, the general lessons of theory and recent experience are summarized.

Use a tax system that taxes realized mineral rent. A mineral tax system that uses resource rent as its tax base is likely to lead to both the optimal pace of extraction and the highest present value of mineral rent to the state, provided that the overall burden of taxation is competitive with that in other countries. Frequent adjustments to taxes are likely to deter investment, cause inefficient resource use, and reduce overall revenues.

Recognize that rewards to foreign capital which reflect international market rates need not mean a national loss of income. The key objective is long-run maximization of revenue from mineral rent. If efficient extraction is best achieved by importing foreign capital and management, insistence on state or national ownership may be a false economy that eventually reduces income. The share of mineral revenues that accrues to the state, and the total over time, is more important than ownership of the plant and equipment.

Use conservative forecasts of mineral revenues. Prediction errors have proved very costly, and making adjustments in the face of faulty predictions is harder to do than boosting spending after a windfall. A 20 to 30 percent chance that prices will be lower than forecast is probably a sufficient risk to take. Surplus balances can be profitably saved abroad.

Spend in the medium term only up to the level that promotes noninflationary growth. Saving abroad is preferable to low-yielding investments at home. High inflation and shortages caused by excessive spending damage preexisting economic activity.

Aim for a stable path in the growth of public expenditure through expected cycles. Fiscal and monetary fine-tuning is difficult (and rejected now in many successful economies). The disruption caused by fluctuating public expenditure is probably more costly than delays in some projects that could be funded.

Separate mineral revenue flows from other revenues and release them to the budget at a steady rate determined through cautious predictions. This is little more than an institutional form of fiscal discipline, but it can help the government resist the buildup of pressure to spend. It is easier to make a case that mineral revenues should be cautiously spent if their fluctuations (and their management) can easily be seen by politicians, administrators, and the public.

Allow adjustment to take place, but choose the mode that offers slowest inflation. When a windfall gain is spent to any extent (and not necessarily with an upswing in a predictable cycle), the real exchange rate appreciates and resources move from less to more profitable activities. If this does not happen, the economy cannot benefit from its comparative advantage. However, the degree to which the shift is beneficial depends on whether the preceding lessons are being observed. In general, it is necessary to "lean against the wind" with some combination of these lessons; otherwise the downside will be grim.

Monitor the labor market and operate a stabilizing-wages policy if there is significant inflexibility. If wages are flexible, if labor supply is relatively elastic, and if aggregate labor supply is not fixed independently of labor market conditions, adjustment can occur at low cost. These conditions are usually not completely fulfilled, so that wage rises and movement of labor in a boom imply significant adjustment costs in a slump, the most important being unemployment. This situation calls for special measures to restrain wage rises in boom periods if the overall stabilization of demand will not do so alone.

Use the opportunity of a windfall to "buy off" special interest groups and relax direct controls on economic activity. The rent-seeking syndrome is a considerable danger, both in the boom when owners of resources in limited supply stand to gain and in the slump if administrative controls on foreign exchange, imports, or prices are in place. Use of windfalls to relax supply constraints and dismantle restrictions (by compensating powerful groups that gain from regulation) may therefore prove fruitful in the long term.

Recognize that the incentive pattern and elimination of structural constraints are as important as investments in capacity. No set of choices on how to spend a windfall is likely to realize the windfall's full potential benefits if the economy is highly distorted and subject to major bottlenecks (such as concentration of land ownership, poor administrative capacity, or inadequate transportation systems). By the same token, good macroeconomic management will not be enough if microeconomic policies in strategic sectors do not permit growth in a boom or recovery in a slump; in low-income countries, policies toward agriculture are especially sensitive.

Confine public-sector spending to the things that governments do (or should do) best and design transfers to the private sector that are as neutral as possible. Public-sector investment is justified where expected social and private returns clearly diverge and where genuine public goods are involved. Otherwise, a government may achieve as much or more by enabling private agents to take the risks and creating a stable environment in which they are likely to do so. This is not an abdication of government initiative, just a different way of taking it.

These lessons offer a counsel of perfection (and one that is not unique to mineral economies). Governments with short time horizons and countries with competing centers of power will continue to operate under the constraints that have contributed to poor performance in the past. Nevertheless, past experience, from African nonfuel mineral exports to "beneficiaries" of the oil booms, has been sufficiently disappointing and the potential for economic and social development is so great that alternative approaches may now be attractive. Governments that have dissipated mineral windfalls seem in any case to have no greater expectation of survival than those that have done better.

Acknowledgment

The author is grateful to the United Kingdom Overseas Development Administration for Research Grant R3848, which supported an earlier research project on a related topic. This paper has been much improved as a result of comments by John Tilton, David Evans, Adrian Wood, Richard Auty, Mike Faber, and an anonymous referee and through discussion with participants in seminars and lectures based on the paper at the Colorado School of Mines, the Institute of Development Studies (University of Sussex), the University of East Anglia, and the House of Commons (All-Party Group on Overseas Development). The author is solely responsible for the contents of this paper.

References

Auty, Richard M. 1989. "The International Determinants of Eight Oil-Exporting Countries' Resource-Based Industry Performance," *Journal of Development Studies* vol. 25, no. 3 (April) pp. 354–372.

Bates, Robert H. 1983. *Essays on the Political Economy of Rural Africa* (Cambridge, England, Cambridge University Press).

Corden, W. M., and J. Peter Neary. 1982. "Booming Sector and Deindustrialisation in a Small Open Economy," *Economic Journal* vol. 92, no. 368 (December) pp. 825–848.

————, and Paul Baxter. 1983. *Exchange Rate and Macro-economic Policy in Papua New Guinea* (Canberra, Australia, Australian National University Press).

Garnaut, Ross, and Anthony Clunies Ross. 1983. *Taxation of Mineral Rents* (Oxford, England, Clarendon Press).

Gelb, Alan, and Associates. 1988. *Oil Windfalls: Blessing or Curse?* (Oxford, England, Oxford University Press).

Gregory, R. G. 1976. "Some Implications of the Growth of the Mineral Sector," *Australian Journal of Agricultural Economics* vol. 20 (August) pp. 71–91.

Hirschman, Albert O. 1958. *The Strategy of Economic Development* (New Haven, Conn., Yale University Press).

International Monetary Fund. 1987. *International Financial Statistics*, December (Washington, D.C., International Monetary Fund).

Krueger, Anne O. 1974. "The Political Economy of the Rent-Seeking Society," *American Economic Review* vol. 64, no. 3 (June) pp. 291–303.

Lewis, Stephen R. 1984. "Development Problems of the Mineral-Rich Countries," in M. Syrquin, L. Taylor, and L. E. Westphal, eds., *Economic Structure and Performance: Essays in Honour of Hollis B. Chenery* (San Diego, Calif., Academic Press).

Lipton, Michael. 1977. *Why Poor People Stay Poor* (London, Temple Smith).

Meade, J. E. 1951. "The Theory of International Economic Policy," vol. 1, *The Balance of Payments* (Oxford, England, Royal Institute of International Affairs).

Myrdal, Gunnar. 1957. *Economic Theory and Underdeveloped Regions* (London, Duckworth).

Nankani, Gobind. 1979. "Development Problems of Mineral Exporting Countries," World Bank Staff Working Paper 354, August (Washington, D.C., World Bank).

Neary, J. Peter, and Sweder Van Wijnbergen, eds. 1986. *Natural Resources and the Macroeconomy* (Cambridge, Mass., MIT Press).

Prebisch, Raul. 1963. *Towards a Dynamic Development Policy for Latin America* (New York, United Nations).

Radetzki, Marian. 1985. *State Mineral Enterprises in Developing Countries: Their Impact on International Mineral Markets* (Washington, D.C., Resources for the Future).

Roemer, Michael. 1985. "Dutch Disease in Developing Countries: Swallowing Bitter Medicine," in Mats Lundahl, ed., *The Primary Sector in Economic Development* (London, Croom Helm).

Salter, W. E. G. 1959. "Internal and External Balance: The Role of Price and Expenditure Effects," *Economic Record* vol. 35 (August) pp. 226–238.

Singer, Hans W. 1950. "The Distribution of Gains Between Investing and Borrowing Countries." *American Economic Review, Papers and Proceedings* vol. 40, no. 2 (May) pp. 473–485.

———. 1984. "Terms of Trade Controversy and the Evolution of Soft Financing: Early Years in the UN, 1947–51," in Gerald M. Meier and Dudley Seers, eds., *Pioneers in Development* (Oxford University Press for the World Bank).

Snape, R. H. 1977. "Effects of Mineral Development on the Economy," *Australian Journal of Agricultural Economics* vol. 21 (December) pp. 146–156.

Tilton, John E. 1981. "Cyclical Instability: A Growing Threat to Metal Producers and Consumers," *Natural Resources Forum* vol. 5, no. 1 (January) pp. 5–14.

United Nations. 1982. *Fluctuations in Primary Commodity Prices and Their Effects on Developing Countries* (New York, United Nations, Department of International Economics and Social Affairs).

Van Wijnbergen, Sweder. 1984. "The Dutch Disease: A Disease After All?" *Economic Journal* vol. 94, no. 373 (May) pp. 233–250.

Wheeler, David. 1984. "Sources of Stagnation in Sub-Saharan Africa," *World Development* vol. 12, no. 1 (January) pp. 1–23.